T0136969

Human Behavior Analysis: Sensing and Understanding

Zhiwen Yu • Zhu Wang

Human Behavior Analysis: Sensing and Understanding

Zhiwen Yu
School of Computer Science
Northwestern Polytechnical University
Xi'an, China

Zhu Wang
School of Computer Science
Northwestern Polytechnical University
Xi'an, China

ISBN 978-981-15-2111-9   ISBN 978-981-15-2109-6   (eBook)
https://doi.org/10.1007/978-981-15-2109-6

This Springer imprint is published by the registered company Springer Nature Singapore Pte Ltd.
The registered company address is: 152 Beach Road, #21-01/04 Gateway East, Singapore 189721, Singapore

Preface

In recent years, human behavior sensing and understanding attracts a lot of interests due to various societal needs, including security, natural interfaces, gaming, affective computing, and assisted living. However, accurate detection and recognition of human behavior is still a big challenge that attracts a lot of research efforts.

Traditionally, to identify human behaviors, we first need to continuously collect the sensory data from physical sensing devices (e.g., camera, GPS, and RFID), which can be either worn by humans, attached on objects, or deployed in environments. Afterwards, based on recognition algorithms or classification models, the behavior types can be identified so as to facilitate upper layer applications. Although such traditional behavior identification approaches perform well and are widely adopted, most of them are intrusive and require specific sensing devices, raising issues such as privacy and deployment cost.

In this monograph, we aim to provide an overview of recent research progress on noninvasive human behavior sensing and understanding. Specifically, this monograph differs from existing literature in the following aspects. On the one hand, we mainly focus on human behavior understanding approaches that are based on noninvasive sensing technologies, including both sensor-based and device-free approaches. On the other hand, while most existing studies are about individual behaviors, we will systematically elaborate how to understand human behaviors of different granularities, including not only individual-level behaviors but also group-level and community-level behaviors.

The book includes four parts. In Part I (Chaps. 1 and 2), we introduce and analyze the design, implementation, and development of a typical human behavior sensing and understanding system and then give the main steps of such a system. Part II (Chaps. 3 and 4) mainly focuses on two noninvasive (i.e., sensor-based and device-free) behavior sensing approaches. In Part III (Chaps. 5–7), we elaborate our studies on the understanding of different granularity human behaviors, from individual level to group level and community level. Finally, in Part IV (Chap. 8), we discuss the open issues and possible solutions involved in human behavior sensing and understanding, followed by a conclusion to the whole monograph. Specifically, some of the contents in this monograph might be of particular interest to readers, including noninvasive human behavior sensing approaches (i.e., sensor-based and device-free), as well as the understanding of different granularity human behaviors (i.e., individual level, group level, and community level).

We would like to thank Prof. Daqing Zhang at the Software Engineering Institute of Peking University, Beijing, China; Prof. Liming Chen at the School of Computer Science and Informatics of De Montfort University, Leicester, UK; Prof. Xingshe Zhou at the School of Computer Science of Northwestern Polytechnical University, Xi'an, China; and Prof. Bin Guo at the School of Computer Science of Northwestern Polytechnical University, Xi'an, China. We would like to thank all of the members of Ubiquitous Computing group of Northwestern Polytechnical University, China, for their valuable discussions, insights, and helpful comments. We would also like to thank the staff at Springer, Ms. Celine Chang and Ms. Jane Li, for their kind help throughout the publication and preparation processes of the monograph.

Xi'an, China   Zhiwen Yu
Xi'an, China   Zhu Wang

Contents

Chapter 1
Introduction

Abstract Human behavior sensing and understanding has been a popular research area during the past decades, which plays an important role in human-computer interaction and public security. It aims to understand what people are doing by sensing and recognizing their activities and their environments. However, accurate detection and recognition of human behavior is still a big challenge that attracts a lot of research efforts. In this chapter, we aim to present an overview of human behavior sensing and understanding techniques, including from vision-based to sensor-based and device-free behavior sensing, from individual to group and community behavior understanding, and from pattern-based to model-based behavior understanding.

Behavior recognition technology has been applied to a number of key applications, including pervasive and mobile computing [1, 2], surveillance-based security [3–10], context-aware computing [11–13], and ambient assistive living [14–18]. Human behavior sensing also serves as the key component in several consumer products. For example, game consoles such as the Nintendo Wii and the Microsoft Kinect rely on the recognition of gestures or even full-body movements to deliver outstanding game interaction with the users. While originally developed for the entertainment sector, these systems have found additional applications, such as for personal fitness training and rehabilitation, and also stimulated new behavior recognition research [19]. Finally, some sports products such as the Philips DirectLife or the Nike+ running shoes integrate motion sensors and offer both amateur and professional athletes feedback on their performance. Moreover, monitoring daily behavior to support medical diagnosis, for rehabilitation, or to assist patients with chronic impairments has been shown to provide key enhancements to traditional medical methods. Some early assistance to encourage humans to adopt a healthy lifestyle was regarded as another important goal, leading to a vast exploration of related human

Part of this chapter is based on a previous work: L. Chen, J. Hoey, C. D. Nugent, D. J. Cook and Z. Yu, "Sensor-Based Activity Recognition," in IEEE Transactions on Systems, Man, and Cybernetics, Part C (Applications and Reviews), vol. 42, no. 6, pp. 790–808, Nov. 2012. DOI: https://doi.org/10.1109/TSMCC.2012.2198883

behaviors, for example, brushing teeth [20] or handwashing, food [21, 22] and medication intake [23, 24], or transportation routines [25].

1.1 From Vision-Based to Sensor-Based and Device-Free Behavior Sensing

1.1.1 Vision-Based Human Behavior Sensing and Recognition

Using vision to sense and understand the behavior of human beings is a very important topic. Vision-based human behavior recognition is the process of labeling image sequences with behavior labels. It is based on the use of visual sensing facilities, such as video cameras, to monitor user behaviors and environmental changes. The generated sensor data are video sequences or digitized visual data. The approaches exploit computer vision techniques, including feature extraction, structural modeling, movement segmentation, action extraction, and movement tracking, to analyze visual observations for pattern recognition.

Specifically, the vision-based behavior recognition methods mainly consist of two steps: representation and classification. Generally, for representation approaches, related literatures follow a research trajectory of global representations, local representations, and recent depth-based representations. Earlier studies attempted to model the whole images or silhouettes and represent human activities in a global manner. The approach in ref. 26 is an example of global representation in which space-time shapes are generated as the image descriptors. Then, the emergence of space-time interest points (STIPs) [27] triggered significant attention to a new local representation view that focuses on the informative interest points. Meanwhile, local descriptors such as histogram of oriented gradients (HOG) and histogram of optical flow (HOF) oriented from object recognition are widely used or extended to 3D in human behavior recognition area. With the upgrades of camera devices, especially the launch of RGBD cameras in the year 2010, depth image-based representations have been a new research topic and have drawn growing concern in recent years. For example, Jalal et al. [28] propose novel multi-fused features for online human behavior recognition system that recognizes human behaviors from continuous sequences of depth map. Yang et al. [29] propose a general scheme of super normal vector (SNV) to aggregate the low-level polynormals into a discriminative representation, which can be viewed as a simplified version of the Fisher kernel representation.

On the other hand, classification techniques keep developing in step with machine learning methods. In fact, lots of classification methods were not originally designed for human behavior recognition. Generally speaking, most behavior classification algorithms can be divided into three categories, namely template-based approaches, generative models, and discriminative models. Template-based approach is a relatively simple and well-accepted approach; however, it can be sometimes computationally expensive. Generative models learn a model of the joint probability P (X, Y) of the inputs X and the label Y, and then P(Y|X) is calculated using algorithms (e.g., the Bayes classifier) to pick the most likely label Y [30]. In contrast, discriminative models determine the result label directly. Typical algorithms of generative models are hidden Markov model (HMM) and dynamic Bayesian network (DBN), while support vector machine (SVM), relevance vector machine (RVM), and artificial neural network (ANN) are typical discriminative models. In recent years, deep learning methods are developed for a large amount of image classification and have achieved good performance. Basically, the deep learning architectures can be categorized into four groups, namely deep neural networks (DNNs), convolutional neural networks (ConvNets or CNNs), recurrent neural networks (RNNs), and some emergent architectures [31]. The ConvNets is the most widely used one among the mentioned deep learning architectures. Compared with traditional machine learning method and their handcrafted features, the ConvNets can learn some representational features automatically [32]. Unlike ConvNets, DNNs still use handcrafted features instead of automatically learning features by deep networks from raw data. RNNs are designed for sequential information. Activity itself is a kind of time-series data and it is a natural thought to use RNNs for activity recognition.

However, it is worth mentioning that while visual monitoring is intuitive and information rich, the vision-based approaches can only work with line-of-sight (LoS) coverage in rich-lighting environments, and also may cause privacy concerns.

1.1.2 Sensor-Based Human Behavior Sensing and Recognition

Sensor-based behavior recognition is based on the use of emerging sensor network technologies for activity sensing and understanding [33–35]. The generated sensor data are mainly time series of state changes and/or various parameter values that are usually processed through data fusion, probabilistic, or statistical analysis methods and formal knowledge technologies for activity recognition. Generally, sensors can be attached to an actor under observation, namely wearable sensors or smartphones, or objects that constitute the activity environment, namely dense sensing.

Preliminary studies have shown that commodity smartphones equipped with accelerometers can also be used for behavior recognition, which are widely available to the general public [36]. Wearable sensors often use inertial measurement units or radio frequency identification (RFID) tags to gather behavioral information. This approach is effective to recognize physical movements such as exercises. In contrast, dense sensing infers activities by monitoring human–object interactions through the usage of multimodal miniaturized sensors. These sensors are different in types, purposes, output signals, underpinning theoretical principles, and technical infrastructure, and can be classified into two main categories in terms of the way they are deployed in activity monitoring applications, i.e., wearable sensors and dense sensors.

Wearable sensors, generally, refer to sensors that are positioned directly or indirectly in human body. They generate signals when the user performs activities. As a result, they can monitor features that are descriptive of the person's physiological state or movement. Wearable sensors can be embedded into clothes, eyeglasses, belts, shoes, wristwatches, and mobile devices or positioned directly in the body. They can be used to collect information such as body position and movement, pulse, and skin temperature. However, wearable sensors are not suitable for monitoring activities that involve complex physical motions and/or multiple interactions with the environment.

In some cases, sensor observations from wearable sensors alone are not sufficient to differentiate activities involving simple physical movements (e.g., making tea and making coffee). As a result, dense sensing-based activity monitoring has emerged. Dense sensing-based activity monitoring refers to the practice that sensors are attached to objects and activities are monitored by detecting user–object interactions. The approach is based on real-world observations that activities are characterized by the objects that are manipulated during their performance. A simple indication of an object being used can often provide useful clues about the behavior being undertaken. As such, it is assumed that activities can be recognized from sensor data that monitors human interactions with objects in the environment. By dense sensing, we refer to the way and scale with which sensors are used. Using dense sensing, a large number of sensors, normally of low cost low power, and miniaturized, are deployed in a range of objects or locations within an environment for the purpose of monitoring movements and behaviors.

1.1.3   Device-Free Human Behavior Sensing and Recognition

Compared with sensor-based behavior sensing, device-free behavior sensing is also an important way to sense the activity of human beings. It can sense the behavior of human beings without the need of carrying any tags or devices. It has the additional advantage of being unobtrusive while offering good privacy protection. Over the past decades, researchers have studied ways of tracking device-free human subjects using different techniques such as camera [37], capacitance [38], pressure [39], infrared [40], and ultrasonic [41]. However, most of these approaches suffer from serious limitations such as occlusion [37, 40], high deployment cost [38, 39], or short range [41].

Radio frequency (RF)-based techniques have the advantages of long range, low cost, and ability to work through nonconducting walls and obstacles. Therefore, RF-based device-free systems become popular to detect, locate, and track human behaviors. Specifically, RF devices include ZigBee, Wi-Fi, RFID, etc. The basic idea of RF-based device-free sensing system is as follows: when located in indoor environment with such a system, the movement of human bodies will affect the wireless signals and change the multipath profile of the system. Based on this principle, we are able to recognize human behaviors by exploring the changes of wireless signals caused by user movements. Especially, Wi-Fi signal-based human activity recognition technologies can be applied at commercial off-the-shelf (COTS) devices. By monitoring the wireless channel state, the receiver can measure small signal changes caused by human movements and use these changes to recognize the surrounding human activities [42].

Device-free behavior sensing technology can be useful in many practical applications including intrusion detection and tracking for home and office applications, which could enhance the safety of law-enforcement personnel, and low-cost long-range asset protection, e.g., border protection or protection of railroad tracks. In addition, device-free behavior sensing technology can be used to enhance traditional security systems, such as motion detection and video surveillance by providing non-line-of-sight (NLoS) detection and lower deployment cost.

1.2   From Individual to Group and Community Behavior Recognition

In the past decades, numerous research efforts have been made to model and recognize the behavior of individuals. However, group and community behavior recognition has attracted much attention recently, as it has many practical applications such as abnormal group detection, group affective computing, video surveillance, and public security [43]. Specifically, on the one hand, the large deployment of sensor network in public facilities, private buildings and outdoors environments, and digital traces left by people while interacting with cyber-physical spaces are accumulating at an unprecedented breadth, depth, and scale. On the other hand, the recent explosion of sensor-rich smartphone market and the phenomenal growth of geo-tagged data (e.g., Twitter messages, Foursquare check-ins) have enabled the analysis of new dimensions of contexts that involve the social and urban context. We call all those traces left by people the "digital footprints." Leveraging the capacity to collect and analyze the "digital footprints" at group or community scale, a new research field called "social and community intelligence (SCI)" [44] is emerging that aims at revealing the patterns of individual, group, and societal behaviors.

The scale and heterogeneity of the multimodal, mixed data sources present us an opportunity to compile the digital footprints into a comprehensive picture of individual's daily life facets, and radically change the way we build computational models of human behaviors. Numerous innovative services will be enabled, including human health, public safety, urban planning, environment monitoring, and so on. The development of social and community intelligence will greatly expand the scale and depth of context-aware computing, from merely personal awareness to the understanding of social interactions (e.g., social relations, community structures) and urban dynamics (e.g., traffic jams, hot spots in cities). For example, Su et al. [45] explored coherent LSTM to model the nonlinear characteristics in crowd behaviors. Zhuang et al. [46] proposed an end-to-end deep architecture, differential recurrent convolutional neural networks (DRCNN), for group activity recognition.

Specifically, people constantly participate in social activities to interact with others and form various communities. Social activities such as making new friends, forming an interest group to exchange ideas, and sharing knowledge with others are constantly taking place in human society. The analysis of social interactions has been studied by social scientists and physicists for a couple of decades [47]. An excellent introduction to the concepts and the mathematical tools for social networks analysis can be referred [48]. In the early stage, efforts on social network analysis are mostly based on the relational data obtained by survey. During the last two decades, we have observed an explosive growth of social applications such as chatting, shopping, and experience sharing. These applications, along with traditional email and instant messaging, have changed the way we communicate with each other and form social communities. Corresponding to this trend, a large amount of work on social network analysis and knowledge discovery springs up, including email communication networks [49], scientific collaboration, and co-authorship network [50].

More recently, as the Internet stepped into the era of Web 2.0, which advocates that users interact with each other as contributors to the websites' content, researchers turned their attention to the online social utilities, such as Facebook, Twitter, and Blogs. For example, ArterMiner [51] seeks to harvest personal profile information from a user's homepage. AmitSheth's research group has done much work on summarization of event information like space, time, and theme from social web resources for building public services [52]. Twitter has been reported to support real-time mining of natural disasters such as earthquakes [53] and the moods of citizens [54]. Understanding human movement in urban environments has direct implications for the design of public transport systems (e.g., more precise bus scheduling, improved services for public transport users), traffic forecasting (e.g., hotspot prediction), and route recommendation (e.g., for transit-oriented urban development). A number of studies have extracted citywide human mobility patterns using large-scale data from smart vehicles, mobile phones, and smart cards used in public transportation systems. The Real Time Rome project of MIT uses aggregated data from buses and taxis to better understand urban dynamics in real time [55]. The learned human mobility patterns are also useful for urban planning. For example, Nicholson and Noble [56] have studied how to leverage the learned human movement dynamics to improve the distribution of cell infrastructure (e.g., to have a better load balance among cell towers) in wireless communication networks. Zhang et al. [57] has put forward a novel cross-modal representation learning method that uncovers urban dynamics with massive geo-tagged social media data.

Specifically, we will mainly focus on individual, group, and community behavior sensing and understanding in Chaps. 5–7, respectively.

1.3 From Pattern-Based to Model-Based Behavior Recognition

Although primitive sensor data can be obtained through behavior monitoring, recognition models are critical to interpret the sensor data to understand human behaviors. In particular, the mechanisms through which behaviors are recognized are closely related to the nature and representation of behavior recognition models. Generally speaking, recognition models can be built using two different approaches, i.e., pattern based and model based [58–60].

1.3.1 Pattern-Based Behavior Recognition

The first one is to learn recognition models from existing large-scale datasets of human behaviors based on data mining and machine learning techniques. This method involves the creation of probabilistic or statistical behavior recognition models, followed by training and learning processes. To build a human behavior sensing and recognition system, the connection between signal variations and human activities must be established. If the signal variation patterns have unique and consistent relations with certain human activities, it is possible for a pattern-based (or learning-based) method to recognize human behaviors accurately from signal patterns. As this method is driven by data, and the behavior inference is based on probabilistic or statistical classification, it is often referred to as data-driven or pattern-based approaches.

The key to designing pattern-based approaches is to observe and find discriminative patterns to construct features and differentiate different human behaviors of interest. The features can be very simple or sophisticated, depending on the complexity of the recognition task and the required granularity. For simple sensing tasks, feature selection is often based on intuition or direct observation. When the number of behaviors that need to be distinguished is small, it is often easy to find regular but differentiable patterns. In this case, one or two features may be enough to distinguish among behaviors. As both the number of human behaviors and the sensing granularity increase, it becomes challenging to find one-to-one mappings between behaviors and feature patterns. One or two simple features are not enough for this task anymore. In this case, more features are needed to increase the dimension of feature space, and more powerful classifiers are needed.

The advantages of data-driven approaches are the capabilities of handling uncertainty and temporal information. However, this method requires large datasets for training and learning, and suffers from the data scarcity or the "cold start" problem. It is also difficult to apply learnt recognition models from one person to another. As such, this method suffers from the problems of scalability and reusability. Despite these drawbacks, pattern-based approaches have been very popular and successful in human behavior sensing and understanding applications because they are not only conceptually intuitive but also relatively simple to design, for both data collection and algorithm development.

1.3.2 Model-Based Behavior Recognition

Different from pattern-based approaches, which often involve nontrivial training effort and could only recognize a limited set of pre-defined behaviors, model-based recognition approaches are based on the understanding and abstraction of a mathematical relationship among human behaviors and the sensory data. Model-based approaches usually exploit rich prior knowledge in the domain of interest to construct models directly using knowledge engineering and management technologies, which involves knowledge acquisition, formal modeling, and representation. Recognition models generated based on this method are normally used for behavior recognition or prediction through formal logical reasoning, e.g., deduction, induction, or abduction. As such, this method is referred to as knowledge-driven or model-based approach.

Knowledge-driven approaches have the advantages of being semantically clear, logically elegant, and easy to get started. However, they are weak in handling uncertainty and temporal information and the models could be viewed as static and incomplete.

References

1. M. Weiser, "The computer for the twenty-first century," Sci Am., vol. 265, no. 3, pp. 94–104, 1991.
2. T. Choudhury, S. Consolvo, and B. Harrison, "The mobile sensing platform: An embedded activity recognition system," IEEE Pervasive Comput., vol. 7, no. 2, pp. 32–41, Apr./Jun. 2008.
3. J.K. Aggarwal and Q. Cai, "Human motion analysis: A review," Comput. Vis. Image Underst., vol. 73, no. 3, pp. 428–440, 1999.
4. C. Cedras and M. Shah, "Motion-based recognition: A survey," Image Vis. Comput., vol. 13, no. 2, pp. 129–155, 1995.
5. D.M. Gavrila, "The visual analysis of human movement: A survey," Comput. Vis. Image Underst., vol. 73, no. 1, pp. 82–98, 1999.
6. R. Poppe, "A survey on vision-based human action recognition," Image Vis. Comput., vol. 28, no. 6, pp. 976–990, 2010.
7. T.B. Moeslund, A. Hilton, and V. Kruger, "A survey of advances in vision-based human motion capture and analysis," Comput. Vis. Image Underst., vol. 104, no. 2, pp. 90–126, 2006.
8. A. Yilmaz, O. Javed, and M. Shah, "Object tracking: A survey," ACM Comput. Surv., vol. 38, no. 4, pp. 1–45, 2006.
9. P. Turaga, R. Chellappa, and O. Udrea, "Machine recognition of human activities: A survey," IEEE Trans. Circuits Syst. Video Technol., vol. 18, no. 11, pp. 1473–1488, Nov. 2008.
10. D. Weinland, R. Ronfard, and E. Boyer, "A survey of vision-based methods for action representation, segmentation and recognition," Comput. Vis. Image Underst., vol. 115, no. 2, pp. 224–241, 2011.

11. K. Laerhoven and K. A. Aidoo, "Teaching context to applications," J. Pers. Ubiquitous Comput., vol. 5, no. 1, pp. 46–49, 2001.
12. C. R. Wren and E. M. Tapia, "Toward scalable activity recognition for sensor networks," in Proc. 2nd Int. Workshop Location Context Awareness, 2006, pp. 168–185.
13. M. Stikicand and B. Schiele, "Activity recognition from sparsely labeled data using multi-instance learning," in Proc. Location Context Awareness, 2009, vol. 5561, pp. 156–173.
14. M. Philipose, K.P. Fishkin, M. Perkowitz, D.J. Patterson, D. Fox, H. Kautz, and D. Hahnel, "Inferring activities from interactions with objects," IEEE Pervasive Comput., vol. 3, no. 4, pp. 50–57, Oct./Dec. 2004.
15. D. Cook and M. Schmitter-Edgecombe, "Assessing the quality of activities in a smart environment," Methods Inf. Med., vol. 48, no. 5, pp. 480–485, 2009.
16. T. Kasterenand and B. Krose, "Bayesian activity recognition in residence for elders," in Proc. Int. Conf. Intell. Environ., Feb. 2008, pp. 209–212.
17. L. Chen and C. D. Nugent, "Ontology-based activity recognition in intelligent pervasive environments," Int. J. Web Inf. Syst., vol. 5, no. 4, pp. 410–430, 2009.
18. G. Singla, D. Cook, and M. Schmitter-Edgecombe, "Recognizing independent and joint activities among multiple residents in smart environments," J. Ambient Intell. Humaniz Comput., vol. 1, no. 1, pp. 57–63, 2010.
19. J. Sung, C. Ponce, B. Selman, and A. Saxena. 2011. Human activity detection from RGBD images. In Proceedings of the AAAI Workshop on Plan, Activity, and Intent Recognition.
20. Jonathan Lester, Tanzeem Choudhury, and Gaetano Borriello. 2006. A practical approach to recognizing physical activities. In Proceedings of the International Conference on Pervasive Computing. 1–16.
21. Oliver Amft, Martin Kusserow, and Gerhard Troster. 2007. Probabilistic parsing of dietary activity events. In Proceedings of BSN. Springer, 242–247.
22. G. Pirkl, K. Stockinger, K. Kunze, and P. Lukowicz. 2008. Adapting magnetic resonant coupling based relative positioning technology for wearable activity recognition. In Proceedings of ISWC. 47–54.
23. D. Wan. 1999. Magic medicine cabinet: A situated portal for consumer healthcare. In Handheld and Ubiquitous Computing. Springer, 352–355.
24. R. de Oliveira, M. Cherubini, and N. Oliver. 2010. MoviPill: Improving medication compliance for elders using a mobile persuasive social game. In Proceedings of UbiComp, Vol. 1001. 36.
25. J. Krumm and E. Horvitz. 2006. Predestination: Inferring destinations from partial trajectories. In Proceedings of UbiComp. 243–260.
26. M. Blank, L. Gorelick, E. Shechtman, M. Irani, and R. Basri, "Actions as space-time shapes," in Tenth IEEE International Conference on Computer Vision (ICCV'05) Volume 1, pp. 1395–1402, Beijing, China, 2005.
27. I. Laptev and T. Lindeberg, "Space-time interest points," in Proceedings Ninth IEEE International Conference on Computer Vision, vol. 1, pp. 432–439, Nice, France, 2003.
28. A. Jalal, Y. Kim, Y. Kim, et al. Robust human activity recognition from depth video using spatiotemporal multi-fused features. Pattern recognition, 2017, 61: 295–308.
29. X. Yang and Y. Tian. Super normal vector for human activity recognition with depth cameras. IEEE transactions on pattern analysis and machine intelligence, 2017, 39(5): 1028–1039.
30. A. Jordan, "On discriminative vs. generative classifiers: a comparison of logistic regression and naive Bayes," Advances in neural information processing systems, vol. 14, p. 841, 2002.
31. S. Min, B. Lee, and S. Yoon, "Deep learning in bioinformatics," Briefings in Bioinformatics, vol. 17, 2016.
32. G. Luo, S. Dong, K. Wang, and H. Zhang, "Cardiac left ventricular volumes prediction method based on atlas location and deep learning," in 2016 IEEE International Conference on Bioinformatics and Biomedicine (BIBM), pp. 1604–1610, Shenzhen, China, 2016
33. A. Pantelopoulos and N. G. Bourbakis, "A survey on wearable sensor-based systems for health monitoring and prognosis," IEEE Trans. Syst., Man, Cybern. C, Appl. Rev., vol. 40, no. 1, pp. 1–12, Jan. 2010.

34. H. Alemdar and C. Ersoy, "Wireless sensor networks for healthcare: A survey," Comput. Netw., vol. 54, no. 15, pp. 2688–2710, 2010.
35. D. Ding, R. A. Cooper, P. F. Pasquina, and L. Fici-Pasquina, "Sensor technology for smart homes," Maturitas, vol. 69, no. 2, pp. 131–136, 2011.
36. Y. Lu, Y. Wei, L. Liu, et al. Towards unsupervised physical activity recognition using smartphone accelerometers. Multimedia Tools and Applications, 2017, 76(8): 10701–10719.
37. J. Krumm, S. Harris, B. Meyers, B. Brumitt, M. Hale, and S. Shafer. Multi-camera multi-person tracking for easy living. In Proceedings of Third IEEE International Workshop on Visual Surveillance, IWVS' 00, pages 3–10, 2000.
38. M. Valtonen, J. Maentausta, and J. Vanhala. Tiletrack: Capacitive human tracking using floor tiles. In IEEE International Conference on Pervasive Computing and Communications, PerCom'09, pages 1–10, 2009.
39. R. J. Orr and G. D. Abowd. The smart floor: a mechanism for natural user identification and tracking. In CHI '00 extended abstracts on Human factors in computing systems, CHI EA '00, pages 275–276, 2000.
40. D. De, W. Song, M. Xu, C. Wang, D. Cook, and X. Huo. Findinghumo: Real-time tracking of motion trajectories from anonymous binary sensing in smart environments. In Proceedings of the 32nd IEEE International Conference on Distributed Computing Systems, ICDCS'12, pages 163–172, 2012.
41. T.W. Hnat, E. Griffiths, R. Dawson, and K. Whitehouse. Doorjamb: unobtrusive room-level tracking of people in homes using doorway sensors. In Proceedings of the 10th ACM Conference on Embedded Network Sensor Systems, SenSys'12, pages 309–322, 2012.
42. Wang W, Liu A X, Shahzad M, et al. Device-free human activity recognition using commercial WiFi devices[J]. IEEE Journal on Selected Areas in Communications, 2017, 35(5): 1118–1131.
43. Li T, Chang H, Wang M, et al. Crowded scene analysis: A survey IEEE transactions on circuits and systems for video technology, 2015, 25(3): 367–386.
44. Zhang, D., Guo, B., & Yu, Z. (2011). The emergence of social and community intelligence. IEEE Computer, 44(7), 21–28.
45. Su H, Dong Y, Zhu J, et al. Crowd Scene Understanding with Coherent Recurrent Neural Networks[C]. IJCAI. 2016, 1: 2.
46. Zhuang N, Yusufu T, Ye J, et al. Group activity recognition with differential recurrent convolutional neural networks[C]. Automatic Face & Gesture Recognition (FG 2017), 2017 12th IEEE International Conference on. IEEE, 2017: 526–531.
47. Freeman, L. C. (2004). The development of social network analysis: A study in the sociology of science. Empirical Press.
48. Wasserman, S., & Faust, K. (1994). Social network analysis: Methods and applications. Cambridge, UK: Cambridge University Press.
49. McCallum, A., Wang, X., & Corrada-Emmanuel, A. (2007). Topic and role discovery in social networks with experiments on Enron and academic email. Journal of Artificial Intelligence Research. 30(1), 249–272.
50. Barabasi, A. L., Jeong, H., Neda, Z., Ravasz, E., Schubert, A., & Vicsek, T. (2002). Evolution of the social network of scientific collaborations. Statistical Mechanics and its Applications, 311 (3–4), 590–614.
51. Tang, J., Jin, R. M., & Zhang, J. (2008). A topic modeling approach and its integration into the random walk framework for academic search. Paper presented at the Meeting of 2008 IEEE International Conference on the Meeting of 2008 IEEE International Conference on Data Mining. Pisa, Italy.
52. Sheth, A. (2010). Computing for human experience – Semantics-empowered sensors, services, and social computing on the ubiquitous web. IEEE Internet Computing, 14 (1), 88–97.
53. Sakaki, T., Okazaki, M., & Matsuo, Y. (2010). Earthquake shakes Twitter users: Real-time event detection by social sensors. Paper presented at the Meeting of WWW 2010 Conference. Raleigh, NC.

54. Bollen, J., Pepe, A., & Mao, H. (2009). Modeling public mood and emotion: Twitter sentiment and socio-economic phenomena. Paper presented at the meeting of WWW 2009 Conference. Madrid, Spain.
55. Calabrese, F., & Ratti, C. (2006). Real time Rome. Networks and Communications Studies, 20 (3–4), 247–258.
56. Nicholson, J., & Noble, B. D. (2008). Bread crumbs: Forecasting mobile connectivity. Paper presented at the Meeting of Mobile Computing and Networking. San Francisco, CA.
57. Zhang C, Zhang K, Yuan Q, et al. Regions, periods, activities: Uncovering urban dynamics via cross-modal representation learning[C]//Proceedings of the 26th International Conference on World Wide Web. International World Wide Web Conferences Steering Committee, 2017: 361–370.
58. Wang Z, Guo B, Yu Z, et al. Wi-Fi CSI-Based Behavior Recognition: From Signals and Actions to Activities. IEEE Communications Magazine, 2018, 56(5): 109–115.
59. Wu D, Zhang D, Xu C, Wang H, Li X (2017) Device-free Wi-Fi human sensing: from pattern-based to model-based approaches. IEEE Communications Magazine, 55(10): 91–97.
60. Chen L, Hoey J, Nugent C D, et al. Sensor-Based Activity Recognition, IEEE Transactions on Systems, Man, and Cybernetics, Part C (Applications and Reviews), 2012, 42(6):790–808.

Chapter 2
Main Steps of Human Behavior Sensing and Understanding

Abstract Human behavior sensing and understanding is a complex process that can be roughly characterized by four main steps, including the following: (1) to choose appropriate sensors to monitor and capture a user's behavior along with the state change of the environment; (2) to collect, store, and process collected information through data analysis techniques and/or knowledge representation formalisms at appropriate levels of abstraction; (3) to extract features in a way that different behaviors can be distinguished accordingly; and (4) to select or develop reasoning/ classification algorithms to infer behaviors from sensor data. For each individual task, a number of methods, technologies, and tools are available for use. It is often the case that the selection of a method used for one task is dependent on the method of another task. We present the comprehensive methods used for each of these steps in the following sections.

2.1 Sensory Data Collection

With the development of sensing technologies and the advanced mobile computing techniques, it becomes a reality that human behavior data not only could be obtained from device-based sensors, but also is accessible from online social networks, as shown in Fig. 2.1. Traditionally, a single device like a GPS tracker is employed to record human trajectories, while it is more convenient and efficient to obtain different aspects of human behavior data since sensors are now integrated and embedded in every portable or wearable device. Specifically, a commercial smartphone now contains a wide range of sensors including accelerometers, gyroscopes, and magnetometers, to name a few, which could be employed to collect human activity data. Besides, the prevalence uses of social network applications, such as Twitter and Facebook, have revealed millions of human behavior records (e.g., check-in trajectories) that could be leveraged to understand human behavior patterns. As a result, gathering behavior data through physical devices and collecting activity trajectories from online social networks are considered as two major domains for human behavior sensing; both domains could provide effective information for research applications.

© Springer Nature Singapore Pte Ltd. 2020   13
Z. Yu, Z. Wang, *Human Behavior Analysis: Sensing and Understanding*,
https://doi.org/10.1007/978-981-15-2109-6_2

Fig. 2.1 Main steps of human behavior sensing and understanding

As a matter of fact, the operating systems running on each device have transformed the hardware sensors into software APIs, and similarly the online social network platforms also provide programming interfaces for developers to access different types of data. Therefore, it is straightforward to collect data from those two distinct domains by applying the following common procedures: researchers or developers first write programs leveraging those available APIs, after that they run the programs in a backend manner, and finally the behavior data are collected and stored for further analysis. For example, web crawler programming could be applied for collecting online human behavior data by leveraging APIs like Twitter APIs [1] and Foursquare APIs [2]. Besides, operating systems like Android and iOS also provide APIs [3, 4] that allow developers to directly access embedded sensor readings in smart devices. Other than this, researchers usually need to recruit volunteers or make the data collection program publicly available, so that they can gather human behavior data through physical devices. However, collecting data from social network platforms does not require such procedures; researchers only have to query corresponding platforms using some indexes, such as a location, a specific word, or a target user ID, to obtain the desired human activity information. By applying the above procedures, one can collect sensory data of human behavior not only from physical devices, but also from online social network platforms.

2.2 Data Preprocessing

The original sensory data, especially from physical sensors, is often incomplete, inconsistent, and likely to contain many errors. For instance, due to the affection on signals from crowd buildings, there might be huge differences between consecutive GPS locations than it should have in real world. Therefore, data preprocessing becomes one sufficient procedure before any insightful analysis can be taken onto the collected data.

In general, the following steps are considered: data anonymization, data cleaning, data filtering, and data transformation. Firstly, data anonymization is required since human behavior is personal sensitive and is usually related to privacy issue, and MD5 hashing [5] and k-anonymity [6] are two popular data anonymization algorithms. After that, data cleaning is applied to deal with problems like data inconsistency and noisy data influence. Specifically, data inconsistency usually appears in data collected from online platforms, where user profile information may contain conflict records. Meanwhile, noisy influence often exists in physical sensory data that sensor readings are affected by other signals or environmental changes, which could cause inaccuracy when monitoring human activities.

Afterwards, researchers have to choose the most effective or meaningful parts for further analysis, which refers to data filtering. In fact, real-world sensory data are usually incomplete, such that human behavior is often recorded as many fragments and only some of them, which contain enough records in terms of timescale or other criterions, are valuable to be analyzed. Hence, data filtering is performed to filter out sensory data by different criterions that could contribute to discovering pattern from those filtered data.

Finally, data transformation is applied to conduct data normalization or aggregation, and data representation before eventually feeding the data into learning models. In particular, different learning models require different data types or formats, such as vector, matrix, and time-series sequences. Therefore, researchers have to transform original data formats into model-orientated formats and normalize the data when needed. After conducting all these above procedures, the raw data is prepared to be imported into learning algorithms or models for further analysis.

2.3   Feature Extraction

Apart from monitoring human behaviors and activities based on physical sensors or cyber information, one significant task is to understand the relationship between activities and sensory data, which aims to discover correlations that could lead to better human interpretations.

Generally, each instance of sensory data consists of several attributes, and feature extraction manages to derive values intended to be informative and nonredundant, which can be used to facilitate the subsequent learning steps. Specifically, since there are various types of sensors, a human activity could be captured from different aspects that result in high-dimensional sensory data. Therefore, some data-driven-based dimensionality reduction methods (e.g., principal component analysis [7]) are employed to extract meaningful features from the original data. Besides, researchers can also define their own features according to expert knowledge. After that, correlation analysis is conducted to discover insightful relationships between extracted features and human activities. Usually, some metrics like Pearson correlation coefficient [8] are applied to measure the relationship between features and activities, while some other metrics like Information Gain [9] are used to determine the significance of different features. By conducting the above procedures, not only features are extracted from the original sensory data, but also some correlations are learned with respect to different tasks and models.

2.4   Human Behavior Modeling and Classification

Many methodologies are proposed to address human behavior modeling and classi-fication problems, and they can be roughly divided into two directions of approaches: logical reasoning and probabilistic reasoning [10]. Specifically, logical reasoning is the process of using a rational, systematic series of steps based on sound mathematical procedures and given statements to arrive at a conclusion. Traditional human activity recognition and modeling usually prefer this approach to infer human behaviors based on deployed sensors' readings, and this kind of method is more like an expert knowledge-based approach. Logical reasoning models are easy to be interpreted and computationally efficient that could be applied in real-time situations.

Compared with logical reasoning, probabilistic reasoning is a pattern-based approach, and is commonly classified into three types: supervised, unsupervised, and semi-supervised learning models. The supervised learning models depend on the labeled dataset by which the parameters could be learned from each category. On the contrast, unsupervised learning models try to find hidden structure in unlabeled data, such as clustering algorithms. Semi-supervised learning falls between the supervised and unsupervised learning, which mainly makes use of both labeled and unlabeled data for training—typically for the sparse of labeled data. Different from logical reasoning, probabilistic reasoning is more robust when facing problems like noisy, uncertain, and incomplete sensory data. However, since it is data-driven-based approach, it is more computationally expensive and difficult to achieve the goal of real-time modeling.

References

1. https://dev.twitter.com/overview/api, visited: 24 Jul 2018.
2. https://developer.foursquare.com/docs/, visited: 24 Jul 2018.
3. https://developer.android.com/reference/, visited: 24 Jul 2018.
4. https://developer.apple.com/documentation/, visited: 24 Jul 2018.
5. https://en.wikipedia.org/wiki/md5, visited: 24 Jul 2018.
6. Sweeney, Latanya. "k-anonymity: A model for protecting privacy." International Journal of Uncertainty, Fuzziness and Knowledge-Based Systems 10.05 (2002): 557–570.
7. Jolliffe, Ian. "Principal component analysis." International encyclopedia of statistical science. Springer, Berlin, Heidelberg, 2011. 1094–1096.
8. Benesty, Jacob, et al. "Pearson correlation coefficient." Noise reduction in speech processing, Springer, Berlin, Heidelberg, 2009. 1–4.
9. Kent, John T. "Information gain and a general measure of correlation." Biometrika 70.1 (1983): 163–173.
10. Liang, Yunji, et al. "Activity Recognition Using Ubiquitous Sensors: An Overview." Wearable Technologies: Concepts, Methodologies, Tools, and Applications. IGI Global, 2018. 199–230.

Chapter 3
Sensor-Based Behavior Recognition

Abstract In this monograph, sensor-based behavior recognition mainly refers to the use of emerging sensor network technologies for behavior monitoring and understanding. The generated sensor data from sensor-based monitoring are mainly time series of state changes and/or various parameter values that are usually processed through data fusion, probabilistic, or statistical analysis methods and formal knowledge technologies for behavior recognition. Specifically, sensors can be attached to an actor under observation, namely wearable sensors or smartphones, or objects that constitute the environment, namely dense sensing. Wearable sensors often use inertial measurement units and radio frequency identification (RFID) tags to gather a user's behavioral information. This approach is effective to recognize physical movements such as physical exercises. In contrast, dense sensing infers behaviors by monitoring human–object interactions through the usage of multiple multimodal miniaturized sensors. In this chapter, we first give a brief introduction to the historical evolution of sensor-based behavior recognition. Afterwards, we present the mobile device-enabled behavior recognition approach, which is a typical type of sensor-based behavior recognition, followed by a discussion on the key issues of developing behavior recognition systems using mobile devices.

3.1 Sensor-Based Behavior Recognition Evolution

The idea of using sensors for behavior monitoring and recognition has been existent since the late 1990s. It was initially pioneered and experimented by the work of the Neural Network house [1] in the context of home automation, and a number of location-based applications aiming to adapt systems to users' whereabouts [2–4]. The approach was soon found to be more useful and suitable in the area of ubiquitous and mobile computing—an emerging area in the late 1990s—due to its

---

Part of this chapter is based on a previous work: L. Chen, J. Hoey, C. D. Nugent, D. J. Cook and Z. Yu, "Sensor-Based Activity Recognition," in IEEE Transactions on Systems, Man, and Cybernetics, Part C (Applications and Reviews), vol. 42, no. 6, pp. 790–808, Nov. 2012. DOI: https://doi.org/10.1109/TSMCC.2012.2198883

© Springer Nature Singapore Pte Ltd. 2020   17
Z. Yu, Z. Wang, *Human Behavior Analysis: Sensing and Understanding*,
https://doi.org/10.1007/978-981-15-2109-6_3

easy deployment. As such, extensive research has been undertaken to investigate the use of sensors in various application scenarios of ubiquitous and mobile computing, leading to considerable work on context awareness [5–7], smart appliances [8, 9], and activity recognition [10–14]. Most research at that time made use of wearable sensors, either dedicated sensors attached to human bodies or portable devices like mobile phones, with application to ubiquitous computing scenarios such as providing context-aware services. Activities being monitored in these researches are mainly physical activities like motion, walking, and running. These early works lay a solid foundation for wearable computing and still inspire and influence today's research.

In the early 2000s, a new sensor-based approach that uses sensors attached to objects to monitor human activities appeared. This approach, which was later dubbed as the "dense sensing" approach, performs activity recognition through the inference of user–object interactions [15, 16]. The approach is particularly suitable to deal with activities that involve a number of objects within an environment, or instrumental activities of daily living (ADL) [17, 18]. Research on this approach has been heavily driven by the intensive research interests and huge research effort on smart home-based assistive living, such as the EU's AAL program [19]. In particular, sensor-based activity recognition can better address sensitive issues in assistive living such as privacy, ethics, and obtrusiveness than conventional vision-based approaches. This combination of application needs and technological advantages has stimulated considerable research activities in a global scale, which gave rise to a large number of research projects, including the House_n [20], CASAS [21], Gator-Tech [22], inHaus [23], AwareHome [24], DOMUS [25], and iDorm [26] projects, to name but a few. As a result of the wave of intensive investigation, there have seen a plethora of impressive works on sensor-based behavior recognition in the past several years [27–40].

While substantial research has been undertaken, and significant progress has been made, the two main approaches, wearable sensor-based and dense sensing-based behavior recognition, are currently still focuses of the research community [41–45]. The former is mainly driven by the ever-popular pervasive and mobile computing, while the latter is predominantly driven by smart environment applications such as AAL. Interests in various novel applications are still increasing and application domains are rapidly expanding.

3.2   Behavior Recognition Based on Mobile Devices

During the past decade, smart mobile devices become more and more popular, which are usually embedded with various sensors, such as accelerometer and GPS. These sensors generate signals when the user performs behaviors. As a result, sensor-enriched mobile devices can monitor features that are descriptive of the person's physiological state or movement, and can be used to collect information such as body position and movement, pulse, and skin temperature. Researchers have found that different types of sensor information are effective to classify different types of activities. For example, accelerometer embedded in smartphones is capable of characterizing human's movements, e.g., standing, walking, and running [46–48]. Similarly, by collecting audio information from the phone's microphone, it is possible to recognize a user's activity, e.g., listening to music, speaking, and sleeping [49, 50], and even monitor running rhythm [51], as well as the sound-related respiratory symptoms of a user, such as sneeze or cough [52]. With various sensor data, including Wi-Fi, accelerometer, compass, and GPS, the user's emotion can also be inferred [53–55].

Generally, mobile device-based behavior recognition systems may operate at multiple scales, enabling applications from personal sensing and group sensing to community sensing [56]. Meanwhile, another key issue is about how much the user (i.e., the person carrying the mobile device) should be actively involved during the sensing activity (e.g., taking the device out of the pocket to collect a sound sample or take a picture). In other words, should the user actively participate, known as participatory sensing, or, alternatively, passively participate, known as opportunistic sensing [57]? Each of these sensing paradigms presents important trade-offs. In this section, we discuss different sensing and understanding scales and paradigms of mobile device-enabled behavior recognition systems.

3.2.1 Behavior Sensing and Understanding Scales

Personal behavior sensing and understanding systems are designed for a single individual. Typical scenarios include tracking the user's exercise routines or automating diary collection. Typically, personal behavior sensing and understanding systems generate data for the sole consumption of the user and are not shared with others. An exception is healthcare systems where limited sharing with medical professionals is common (e.g., primary care giver or specialist).

Individuals who use behavior sensing and understanding systems that share a common goal, concern, or interest collectively represent a group. These group behavior recognition systems are likely to be popular and reflect the growing interest in social networks or connected groups (e.g., at work, in the neighborhood, friends) who may want to share information freely or with privacy protection. There is an element of trust in group behavior sensing and understanding systems that simplify otherwise difficult problems, such as attesting that the collected sensor data is correct or reducing the degree to which aggregated data must protect the individual. Common use cases include assessing neighborhood safety, sensor-driven mobile social networks, and forms of citizen science.

Most examples of community behavior sensing and understanding only become useful once they have a large number of people participating, e.g., tracking the spread of disease across a city, congestion patterns across city roads, or a noise map of a city. These systems represent large-scale data collection and analysis for the good of the community. To achieve scale implicitly requires the cooperation of strangers who will not trust each other. This increases the need for systems with strong privacy protection and low commitment levels from users.

3.2.2   *Behavior Sensing and Understanding Paradigms*

One issue common to the different types of sensing and understanding scale is to what extent the user is actively involved in the system [57]. We discuss two points in the design space: participatory sensing, where the user actively engages in the data collection activity (i.e., the user manually determines how, when, what, and where to sample), and opportunistic sensing, where the data collection stage is fully automated with no user involvement.

The benefit of opportunistic sensing is that it lowers the burden placed on the user, allowing overall participation by a population of users to remain high even if the system is not that personally appealing. This is particularly useful for community behavior sensing and understanding. One of the main challenges of using opportunistic sensing is the context problem of mobile devices, e.g., the system wants to only take a sound sample for a citywide noise map when the device is out of the pocket or bag. These types of context issues can be solved by using the embedded sensors; for example, the accelerometer or light sensors can determine if the device is out of the pocket.

Participatory sensing, which is gaining interest in the mobile crowd sensing community [57], places a higher burden or cost on the user, e.g., manually selecting data to collect (e.g., lowest petrol prices) and then sampling it (e.g., taking a picture). An advantage is that complex operations can be supported by leveraging the intelligence of the person in the loop who can solve the context problem in an efficient manner. For example, a person who wants to participate in collecting a noise or air quality map of their neighborhood simply takes out his/her mobile device to solve the context problem. One drawback of participatory sensing is that the quality of data is dependent on participant enthusiasm to reliably collect sensing data and the compatibility of a person's mobility patterns to the intended goals of the system (e.g., collect pollution samples around schools).

Clearly, opportunistic and participatory represent extreme points in the design space. Each approach has pros and cons, and there is a need to develop models to best understand the usability and performance issues of these schemes. In addition, it is likely that many systems will emerge that represent a hybrid of both these sensing paradigms.

3.3 Energy-Efficient Behavior Recognition Using Ubiquitous Sensors

With the popularity of mobile devices equipped with unprecedented sensing capabilities, context-aware applications on mobile devices are going flourishing. However, the long-term sensing with the full working load of sensors is energy consuming. The battery capacity of mobile devices is a major bottleneck of context-aware applications. For example, the battery lifetime of Samsung i909 reaches up to over 30 h when all applications and sensors are turned off. But that declines to 5.5 h (50 Hz) and 8 h (20 Hz), respectively, when the single 3D accelerometer is monitored with different sampling frequencies.

In the wireless networks, the use of energy harvesting techniques offers a way of supplying sensor systems without the need for batteries and maintenance [58, 59]. For ubiquitous sensors, existing solutions extend the battery life by the collaboration of multiple sensors and the reduction of sensor active time [60]. Wang et al. [61] designed a scalable framework of energy-efficient mobile sensing system for automatic user state recognition. The core component is a sensor management scheme which defines user states and defines transition rules by an XML configuration. The sensor management scheme allocates the minimum set of sensors and invokes new sensors when state transitions happen. Zappi et al. [62] selected the minimum set of sensors according to their contributions to classification accuracy as assessed during data training process and tested this solution by recognizing manipulative activities of assembly-line workers in a car production environment. Li et al. [63] applied machine learning technologies to infer the status of heavy-duty sensors for energy-efficient context sensing. They try to infer the status of high-energy-consuming sensors according to the outputs of lightweight sensors.

Towards the energy efficiency of behavior recognition based on a single sensor (e.g., accelerometer) in the mobile device, it is intuitive to reduce the working time of sensors by adopting a low sampling frequency. Lower sampling frequency means less work time for the heavy-duty sensor. However, whether the low sample frequency is feasible for detecting human behaviors is still an open question. It is claimed that the sampling rate to assess daily physical activities should be no less than 20 Hz [64–66]. Kahatapitiya et al. [67] utilized the harvesting signal to estimate the step count, completely removing the requirement of accelerometer sampling. On the other hand, low sampling frequency may result in the loss of sampling data, reducing the recognition rate with low-resolution sensory data [68]. So there is a trade-off between energy consumption and recognition rate. Furthermore, many classification algorithms are heavyweight and time consuming for mobile devices. The size of sliding window in most classification algorithms is constant, which not only reduces the ability to detect short-duration movements, but also occupies lots of resources with the consumption of battery power.

To overcome above issues, two factors (i.e., sampling frequency and computational load) should be considered in the design of the behavior recognition algorithm. For example, we propose an energy-efficient method to recognize user activities based on a single triaxial accelerometer embedded in smartphones, where a hierarchical recognition scheme is adopted to reduce the probability of time-consuming frequency-domain features for lower computational complexity and adjust the size of sliding window to enhance the recognition accuracy [69].

References

1. M. C. Mozer, "The neural network house: An environment that adapts to its inhabitants," in Proc. AAAI Spring Symp. Intell. Environ., 1998, pp. 110–114.
2. U. Leonhardt and J. Magee, "Multi-sensor location tracking," in Proc. 4th ACM/IEEE Int. Conf. Mobile Comput. Netw., 1998, pp. 203–214
3. A. R. Golding and N. Lesh, "Indoor navigation using a diverse set of cheap, wearable sensors," in Proc. 3rd Int. Symp. Wearable Comput., Oct. 1999, pp. 29–36.
4. A. Ward and A. Hopper, "A new location technique for the active office," IEEE Personal Commun., vol. 4, no. 5, pp. 42–47, Oct. 1997.
5. A. Schmidt, M. Beigl, and H. Gellersen, "There is more to context than location," Comput. Graph., vol. 23, no. 6, pp. 893–901, 1999.
6. C. Randell and H. L. Muller, "Context awareness by analyzing accelerometer data," in Proc. 4th Int. Symp. Wearable Comput., 2000, pp. 175–176.
7. H. W. Gellersen, A. Schmidt, and M. Beigl, "Multi-sensor context awareness in mobile devices and smart artifacts," Mobile Netw. Appl., vol. 7, no. 5, pp. 341–351, Oct. 2002.
8. A. Schmidt and K. Van Laerhoven, "How to build smart appliances," IEEE Pers. Commun., vol. 8, no. 4, pp. 66–71, Aug. 2001.
9. K. Van Laerhoven and K. A. Aidoo, "Teaching context to applications," J. Pers. Ubiquitous Comput., vol. 5, no. 1, pp. 46–49, 2001.
10. K. Van Laerhoven, K. Aidoo, and S. Lowette, "Real-time analysis of data from many sensors with neural networks," in Proc. 5th Int. Symp. Wearable Comput., 2001, pp. 115–123.
11. K. Van Laerhoven and O. Cakmakci, "What shall we teach our pants?" in Proc. 4th Int. Symp. Wearable Comput., 2000, pp. 77–84.
12. F. Foerster and J. Fahrenberg, "Motion pattern and posture: Correctly assessed by calibrated accelerometers," Behav. Res. Methods Instrum. Comput., vol. 32, no. 3, pp. 450–457, 2000.
13. K. Van Laerhoven and H. W. Gellersen, "Spine versus Porcupine: A study in distributed wearable activity recognition," in Proc. 8th Int. Symp. Wearable Comput., 2004, pp. 142–150.
14. S. W. Lee and K. Mase, "Activity and location recognition using wearable sensors," IEEE Pervasive Comput., vol. 1, no. 3, pp. 24–32, Jul.–Sep. 2002.
15. L. Bao and S. Intille, "Activity recognition from user-annotated acceleration data," in Proc. Pervasive, 2004, vol. 3001, pp. 1–17.
16. D. J. Patterson, L. Liao, D. Fox, and H. Kautz, "Inferring high-level behavior from low-level sensors," in Proc. 5th Conf. Ubiquitous Comput., 2003, pp. 73–89.
17. M. Chan, D. Esteve, C. Escriba, and E. Campo, "A review of smart homes—present state and future challenges," Comput. Methods Programs Biomed., vol. 91, no. 1, pp. 55–81, 2008.
18. C. D. Nugent, "Experiences in the development of a smart lab," Int. J. Biomed. Eng. Technol., vol. 2, no. 4, pp. 319–331, 2010.
19. The ambient assisted living joint programme [Online]. Available: www.aal-europe.eu
20. The house of the future [Online]. Available: http://architecture.mit.edu/house_n
21. P. Rashidi and D. Cook, "Activity knowledge transfer in smart environments," J. Pervasive Mobile Comput., vol. 7, no. 3, pp. 331–343, 2011.
22. The gator-tech smart house [Online]. Available: http://www.icta.ufl.edu/gt.htm
23. The inHaus project in Germany [Online]. Available: http://www.inhaus.fraunhofer.de
24. The aware home [Online]. Available: http://awarehome.imtc.gatech.edu

25. The DOMUS laboratory [Online]. Available: http://domus.usherbrooke.ca
26. The iDorm project [Online]. Available: http://cswww.essex.ac.uk/iieg/idorm.htm
27. N. Kern, B. Schiele, H. Junker, P. Lukowicz, and G. Troster, "Wearable sensing to annotate meeting recordings," Pers. Ubiquitous Comput., vol. 7, no. 5, pp. 263–274, Oct. 2003.
28. P. Lukowicz, J. A. Ward, H. Junker, and T. Starner, "Recognizing workshop activity using body worn microphones and accelerometers," in Proc. Pervasive Comput., Apr. 2004, pp. 18–23.
29. J. Mantyjarvi, J. Himberg, and T. Seppanen, "Recognizing human motion with multiple acceleration sensors," in Proc. IEEE Int. Conf. Syst., Man, Cybern., 2001, vol. 2, pp. 747–752.
30. D. Ashbrook and T. Starner, "Using GPS to learn significant locations and predict movement across multiple users," Pers. Ubiquitous Comput., vol. 7, no. 5, pp. 275–286, Oct. 2003.
31. M. Philipose, K. P. Fishkin, M. Perkowitz, D. J. Patterson, D. Fox, H. Kautz, and D. Hahnel, "Inferring activities from interactions with objects," IEEE Pervasive Comput., vol. 3, no. 4, pp. 50–57, Oct./Dec. 2004.
32. K. P. Fishkin, M. Philipose, and A. Rea, "Hands-on RFID: Wireless wearables for detecting use of objects," in Proc. 9th IEEE Int. Symp. Wearable Comput., 2005, pp. 38–43.
33. D. J. Patterson, D. Fox, H. Kautz, and M. Philipose, "Fine-grained activity recognition by aggregating abstract object usage," in Proc. 9th IEEE Int. Symp. Wearable Comput., 2005, pp. 44–51.
34. M. R. Hodges and M. E. Pollack, "An object-use fingerprint: The use of electronic sensors for human identification," in Proc. 9th Int. Conf. Ubiquitous Comput., 2007, pp. 289–303.
35. M. Buettner, R. Prasad, M. Philipose, and D. Wetherall, "Recognizing daily activities with RFID-based sensors," in Proc. 11th Int. Conf. Ubiquitous Comput., 2009, pp. 51–60.
36. D. Wilson and C. Atkeson, "Simultaneous tracking and activity recognition (STAR) using many anonymous, binary sensor," in Proc. 3rd Int. Conf. Pervasive Comput., 2005, pp. 62–79.
37. C. R. Wren and E. M. Tapia, "Toward scalable activity recognition for sensor networks," in Proc. 2nd Int. Workshop Location Context Awareness, 2006, pp. 168–185.
38. R. Aipperspach, E. Cohen, and J. Canny, "Modeling human behavior from simple sensors in the home," in Proc. Pervasive, 2006, vol. 3968, pp. 337–348.
39. T. Gu, Z. Wu, X. Tao, H. K. Pung, and J. Lu, "epSICAR: An emerging patterns based approach to sequential, interleaved and concurrent activity recognition," in Proc. 7th IEEE Int. Conf. Pervasive Comput. Commun., 2009, pp. 1–9.
40. L. Liao, D. J. Patterson, D. Fox, and H. Kautz, "Learning and inferring transportation routines," Artif. Intell., vol. 171, no. 5–6, pp. 311–331, 2007.
41. A. Montanari, C. Mascolo, K. Sailer, and S. Nawaz, "Detecting Emerging Activity-Based Working Traits through Wearable Technology," Proceedings of the ACM on Interactive, Mobile, Wearable and Ubiquitous Technologies, vol. 1, no. 3, pp. 86, 2017.
42. Y. Zhang, M. Haghdan, and K. S. Xu, "Unsupervised motion artifact detection in wrist-measured electrodermal activity data," Proceedings of the ACM on Interactive, Mobile, Wearable and Ubiquitous Technologies, 2017.
43. J. M. Echterhoff, J. Haladjian, and B. Brügge, "Gait and jump classification in modern equestrian sports," In Proceedings of the 2018 ACM International Symposium on Wearable Computers, pp. 88–91, 2018.
44. S. A. Elkader, M. Barlow, and E. Lakshika. "Wearable sensors for recognizing individuals undertaking daily activities." Proceedings of the 2018 ACM International Symposium on Wearable Computers. ACM, 2018.
45. V. Becker, P. Oldrati, L. Barrios, and G. Sörös, "Touchsense: classifying finger touches and measuring their force with an electromyography armband," In Proceedings of the 2018 ACM International Symposium on Wearable Computers, pp. 1–8, 2018.
46. P. Siirtola and J. Röning, "Recognizing human activities user independently on smartphones based on accelerometer data," Int. J. Interact. Multimedia Artif. Intell., vol. 1, no. 5, pp. 38–45, 2012.

47. A. M. Khan, A. Tufail, A. M. Khattak, and T. H. Laine, "Activity recognition on smartphones via sensor-fusion and KDA-based SVMS," Int. J. Distrib. Sensor Netw., vol. 10, no. 5, pp. 1–14, 2014.
48. M. Shoaib, S. Bosch, Ö. D. Incel, H. Scholten, and P. J. M. Havinga, "Fusion of smartphone motion sensors for physical activity recognition," Sensors, vol. 14, no. 6, pp. 10146–10176, 2014.
49. H. Lu, W. Pan, N. D. Lane, T. Choudhury, and A. T. Campbell, "Soundsense: Scalable sound sensing for people-centric applications on mobile phones," in Proc. MobiSys, 2009, pp. 165–178.
50. H. Du, Z. Yu, F. Yi, Z. Wang et al., "Recognition of group mobility level and group structure with mobile devices," IEEE Transactions on Mobile Computing, vol. 17, no. 4, pp. 884–897, 2018.
51. T. Hao, G. Xing, and G. Zhou, "RunBuddy: A smartphone system for running rhythm monitoring," in Proc. UBICOMP 2015, pp. 133–144.
52. X. Sun, Z. Lu, W. Hu, and G. Cao, "Symdetector: Detecting sound-related respiratory symptoms using smartphones," in Proc. UbiComp 2015, pp. 97–108.
53. A. Mottelson and K. Hornbæk, "An affect detection technique using mobile commodity sensors in the wild," In Proceedings of the 2016 ACM International Joint Conference on Pervasive and Ubiquitous Computing, pp. 781–792, 2016.
54. B. Cao, L. Zheng, C. Zhang, P. S. Yu, A. Piscitello, et al. "Deepmood: modeling mobile phone typing dynamics for mood detection," In Proceedings of the 23rd ACM SIGKDD International Conference on Knowledge Discovery and Data Mining," pp. 747–755, 2017.
55. X. Zhang, W. Li, X. Chen, and S. Lu. MoodExplorer: Towards Compound Emotion Detection via Smartphone Sensing. Proceedings of the ACM on Interactive, Mobile, Wearable and Ubiquitous Technologies, vol. 1, no. 4, pp. 176, 2018.
56. Lane N D, Miluzzo E, Lu H, et al. A survey of mobile phone sensing[J]. IEEE Communications Magazine, 2010, 48(9):140–150.
57. B. Guo, Z. Wang, Z. Yu, Y. Wang, N.Y. Yen, R. Huang, and X. Zhou. Mobile Crowd Sensing and Computing: The Review of an Emerging Human-Powered Sensing Paradigm. ACM Comput. Surv. 48, 1, Article 7 (August 2015), 31 pages.
58. K. Ylli, D. Hoffmann, A. Willmann, P. Becker, B. Folkmer, and Y. Manoli. Energy harvesting from human motion: exploiting swing and shock excitations. Smart Materials and Structures, vol. 24, no. 2, pp. 025–029, 2015.
59. K. Li, C. Yuen, B. Kusy, R. Jurdak, A. Ignatovic, S. Kanhere, and S. K. Jha. Fair scheduling for data collection in mobile sensor networks with energy harvesting. IEEE Transactions on Mobile Computing, 2018.
60. D. Liaqat, S. Jingoi, E. de Lara, A. Goel, W. To, et al. Sidewinder: An energy efficient and developer friendly heterogeneous architecture for continuous mobile sensing. ACM SIGARCH Computer Architecture News, vol. 44, no. 2, pp. 205–215, 2016.
61. Wang Y, Lin J, Annavaram M, Quinn JA, Jason H, Bhaskar K, Sadeh N (2009) A framework of energy efficient mobile sensing for automatic user state recognition. In: Proceedings of the 7th ACM international conference on mobile systems, applications, and services, New York, pp 179–192.
62. Zappi P, Lombriser C, Stiefmeier T, Farella E, Roggen D, Benini L, Troster G (2008) Activity recognition from on-body sensors: accuracy-power trade-off by dynamic sensor selection. Wirel Sens Netw Lect Notes Comput Sci 943: 17–33.
63. Li X, Cao H, Chen E, Tian J (2012) Learning to infer the status of heavy-duty sensors for energy efficient context-sensing. ACM Trans Intell Syst Technol 3(2):1–23.
64. Bouten C, Koekkoek K, Verduin M, Kodde R, Janssen JD (1997) A triaxial accelerometer and portable data processing unit for the assessment of daily physical activity. IEEE Trans Biomed Eng 44:136–147.

65. Khan AM, Lee Y, Lee SY, Kim T (2010) A triaxial accelerometer-based physical-activity recognition via augmented-signal features and a hierarchical recognizer. IEEE Trans Inf Technol Biomed 14:1166–1172.

66. Kwapisz JR, Weiss GM, Moore SA (2010) Activity recognition using cell phone accelerometers. ACM SIGKDD Explorations Newsletter 12.

67. K. Kahatapitiya, C. Weerasinghe, J. Jayawardhana, H. Kuruppu, K. Thilakarathna, and D. Días, Low-power step counting paired with electromagnetic energy harvesting for wearables. In Proceedings of the 2018 ACM International Symposium on Wearable Computers. pp. 218–219, 2018.

68. Maurer U, Smailagic A, Siewiorek DP, Deisher M (2006) Activity recognition and monitoring using multiple sensors on different body positions. In: Proceedings of international workshop on wearable and implantable body sensor networks, Cambridge, pp 113–116.

69. Liang Y, Zhou X, Yu Z, et al. Energy-Efficient Motion Related Activity Recognition on Mobile Devices for Pervasive Healthcare. Mobile Networks & Applications, 2014, 19(3):303–317.

Chapter 4
Device-Free Behavior Recognition

Abstract Traditional methods to sense and recognize human behavior include using wearable devices, cameras, and devices embedded in the environment. Recently, a new kind of behavior sensing approach, device-free behavior sensing, attracts a great amount of interests as it holds the promise to provide a ubiquitous sensing solution by using the pervasive signal (including RF signal, acoustic signal, optical signal, etc). In this chapter, we first introduce the basic concept of device-free behavior sensing and understanding, and then present two typical device-free behavior sensing approaches, i.e., Wi-Fi based and acoustic based.

4.1 The Basic Concept of Device-Free Behavior Sensing and Recognition

The basic idea of device-free sensing can be summarized as follows: human behavior in the range where wireless signal that passes will incur signal reflection, scattering, and diffraction, and finally results in changes of the received signal. By exploiting these changes, we can infer specific human behaviors.

Generally speaking, there are mainly three different kind of wireless signals that have been applied to sense human behavior. The first kind of wireless signal is the RF signal, and Wi-Fi sensing is the most attractive RF-based wireless sensing technology. It provides a ubiquitous sensing solution by only using existing Wi-Fi infrastructure. Another kind of wireless signal is the acoustic signal, which can be transmitted and received by both commodity audio device and customized audio devices. Compared with the RF signal, the wavelength of acoustic signal is relatively shorter, making it capable of providing more accurate sensing results. The third kind of wireless signal is the optical signal (e.g., visible light). Generally, transmitting and receiving the optical signal require specific and relatively expensive devices. In addition, the optical signal is usually affected by the ambient lighting conditions.

© Springer Nature Singapore Pte Ltd. 2020

Z. Yu, Z. Wang, *Human Behavior Analysis: Sensing and Understanding*,

https://doi.org/10.1007/978-981-15-2109-6_4

4.1.1   General Methodology

There are two basic methods for wireless human behavior sensing: velocity estimation and distance estimation.

Velocity estimation: Velocity is a key characteristic of human behavior. Different behavior shows different velocity change. Thus, we can sense and recognize human behavior by exploiting the velocity variation over time. Generally, we can estimate the velocity of moving objects using Doppler effect. Specifically, radial velocity estimation requires at least one wireless signal transceiver, 2-dimension velocity estimation requires at least two wireless signal transceivers, and 3-dimension velocity estimation requires at least three wireless signal transceivers. The transceivers have to be kept steady when estimating absolute velocity.

Distance estimation: Real-time distance measurement is another way to sense human behavior. The variation of distances from object to transceivers indicates the trace of object movement. Traditional distance estimation methods include pulse ranging, FMCW ranging, and phase ranging.

4.1.2   Typical Applications

Wireless sensing is capable of enabling different kinds of human behavior sensing and recognition applications, including health monitoring, indoor localization, security monitoring, etc.

- *Health monitoring*. Wireless sensing technology can be applied to monitor human health status. For example, wireless sensing technology can be used for contactless human respiration and heartbeat monitoring during sleeping. In addition, it can also be applied to detect falling and other abnormal behaviors without interfering daily activity.
- *Indoor localization*. Indoor localization is the basis of multiple interesting applications, such as indoor navigation and advertising push. Wireless sensing technique can achieve high-accuracy indoor localization without any wearable devices, and the state-of-the-art positioning accuracy is decimeter scale.
- *Security monitoring*. Existing security monitoring systems are implemented using traditional camera and infrared devices. These light-based systems have two major drawbacks: (1) we have to deploy many devices to cover the target area and (2) it may raise privacy concerns. Wireless sensing technique is capable of fixing the above two problems. Therefore, the security system based on wireless sensing technique is promising to be a satisfactory solution in specific environments.

4.2 Wi-Fi CSI-Based Behavior Sensing and Recognition[1]

Human behavior understanding plays an important role in human-computer interaction and public security. Researchers have developed many methods using cameras, radars, or wearable sensors [1]. However, they also suffer from certain shortcomings (e.g., user privacy, sensing coverage range). For example, the vision-based approaches only work with line-of-sight (LoS) coverage and rich-lighting environments, which also causes privacy concerns. The low-cost 60 GHz radar solutions can only offer an operation range of tens of centimeters, and such radar devices are not widely deployed in our daily living environments. Wearable sensor-based approaches require people to wear some extra devices.

With the recent advances in wireless communications, behavior recognition based on Wi-Fi has been attracting more and more attention due to its ubiquitous availability in indoor areas. Moreover, Wi-Fi-based behavior recognition approach is able to overcome the aforementioned shortcomings of traditional approaches, as it only leverages the wireless communication feature and does not need any physical sensor.

A typical Wi-Fi-based behavior recognition system consists of a Wi-Fi access point (AP) and one or several Wi-Fi-enabled devices in the environment. When located in indoor environment with such a system, the movement of human bodies will affect the wireless signals and change the multipath profile of the system. Based on this principle, we are able to recognize human behaviors by exploring the changes of wireless signals caused by user movements.

Recently, numerous studies have devoted to pervasive sensing using Wi-Fi devices, such as indoor localization [2] and gesture recognition [3]. These studies are mainly based on received signal strength indication (RSSI). However, the RSSI can fluctuate dramatically even at a stationary link, which makes these detection results unreliable. Quite recently, channel state information (CSI), i.e., the fine-grained information regarding Wi-Fi communication, becomes available [4]. Specifically, CSI describes how the signal propagates from the transmitter to the receiver and reflects the combined effects of the surrounding objects (e.g.,' scattering, fading, and power decay with distance). Meanwhile, there are a set of subcarriers in CSI, each of which contains the information of attenuation and phase shift. Therefore, CSI contains rich information and is more sensitive to environmental variances caused by moving objects [5].

CSI is a metric which estimates the channel by representing the channel properties of a wireless communication link. In the frequency domain, the wireless channel can be described as $Y = H \times X + N$, where X and Y correspond to the transmitted and

[1]Part of this section is based on a previous work: Z. Wang, B. Guo, Z. Yu and X. Zhou, "Wi-Fi CSI-Based Behavior Recognition: From Signals and Actions to Activities," in IEEE Communications Magazine, vol. 56, no. 5, pp. 109-115, May 2018. DOI: https://doi.org/10.1109/MCOM.2018.1700144

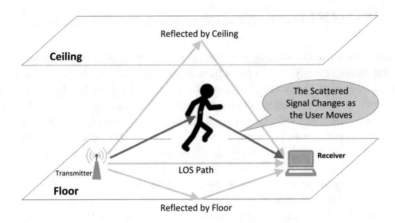

Fig. 4.1 Wi-Fi signal propagation in indoor environments [6]

received signal vectors, H is the channel matrix presented in the format of CSI, and N is the additive white Gaussian noise vector.

In the IEEE 802.11n standard, CSI is measured and reported at the scale of orthogonal frequency division modulation (OFDM) subcarriers, where each $CSI_i = | CSI_i | \exp \{j(\angle CSI_i)\}$ depicts the amplitude response (i.e., $|CSI_i|$) and phase response (i.e., $\angle CSI_i$) of one subcarrier. Specifically, each entry in matrix H corresponds to the channel frequency response (CFR) value between a pair of antennas at a certain OFDM subcarrier frequency at a particular time, and the time series of CFR values for a given pair of antennas and OFDM subcarrier is called a CSI stream. In other words, while CFR describes the combined effects of fading, scattering, and attenuation of a specific subcarrier, CSI is the union of these CFRs. Specifically, 802.11n specifications have provisions for reporting quantized CSI field per packet using various subcarrier grouping options as per clause. However, different manufacturers may choose to implement a subset of the subcarrier grouping options. For example, the Intel 5300 wireless network interface card (NIC) implements an OFDM system with 56 subcarriers of a 20 MHz channel or 114 subcarriers of a 40 MHz channel, 30 out of which can be read for CSI information via the device driver. Thereby, a time series of CSI values includes $30 \times Num_{Tx} \times Num_{Rx}$ CSI streams, where Num_{Tx} and Num_{Rx} stand for the amount of transmitting and receiving antennas, respectively.

Given an indoor environment with two wireless nodes, as shown in Fig. 4.1, the wireless signal will propagate in a multipath manner, and the wireless channel will be relatively stable as long as there are no people or no motion. However, once a person moves, the scattered signals will change (the red line in Fig. 4.1), which causes channel disturbances, involving both amplitude attenuation and phase distortion. In other words, different multipath effects can be obtained if a person is moving, which results in different CSI streams at the receiver, and can be used to recognize different behaviors by correlating them with the corresponding channel distortion patterns.

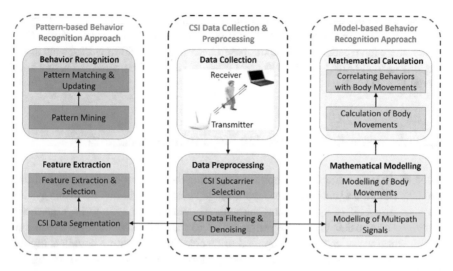

Fig. 4.2 A general architecture of Wi-Fi CSI-based behavior recognition approaches [6]

Wi-Fi CSI-enabled behavior recognition approaches can be categorized into two groups, i.e., pattern based [7–17] and model based [18–21]. The pattern-based approaches aim to classify behaviors by exploring different features of CSI measurements, while the model-based approaches implement recognition by modeling the relationship between signal space and behavior space. A general architecture of Wi-Fi CSI-based behavior recognition approaches is shown in Fig. 4.2. Though the middle part (i.e., CSI data collection and preprocessing) is common to both pattern-based and model-based approaches, the left and right parts illustrate the key difference of these two approaches.

Most existing CSI-based behavior recognition studies adopt the pattern-based approach. The intuition is that different behaviors have distinct impacts on the received CSI streams, which can be leveraged to mine patterns or construct profiles for predefined behaviors, as shown in the left part of Fig. 4.2. Afterwards, each behavior can be classified as one of the predefined types based on profile matching or pattern recognition. The key benefit of the pattern-based recognition approaches is that they do not require intensive deployments and can work with even a single AP, which ensures the low hardware cost and maintenance and has no obstruction to human's normal life. However, pattern-based approaches usually require a learning process to construct profiles or classifiers, which restricts them to identify only a limited set of predefined behaviors.

Model-based approaches are based on the characterization of mathematical relationships between human behaviors and received signals. In the case of Wi-Fi CSI-based behavior recognition, the aim of modeling is to relate the signal space to the physical space including human and environment, and characterize the physical law through mathematical relationship between the received CSI signals and the sensing target, as shown in the right part of Fig. 4.2. Since the model-based approaches do not need predefined behavior profiles, they can track an arbitrary set of human behaviors, which enables wider ranges of real-time applications. Currently, there have been several model-based behavior recognition works, such as the angle-of-arrival (AoA) model [18], the CSI-speed model [19], and the Fresnel zone model [20, 21].

4.3   Acoustic-Based Behavior Sensing and Recognition

Acoustic-based sensing is a method for estimating the state of objects by transmitting acoustic signals and analyzing the response. A typical acoustic-based sensing system consists of a pair of speaker and microphone, which forms an audio transceiver system. The speaker is usually programmed to transmit acoustic signals continuously, while the microphone is applied to receive the echo with a certain sampling rate (e.g., 48 kHz) and send the received signal to computers for data processing and behavior recognition.

Most of the acoustic-based device-free behavior sensing and recognition systems adopt similar ideas from RF-based (e.g., Wi-Fi) approaches, either exploring the Doppler effect when human approaching or away from the microphone, or decoding the echo of frequency-modulated continuous wave (FMCW) of the acoustic signal to measure the human body, or utilizing the OFDM to achieve real-time behavior tracking, as shown in Fig. 4.3.

The first line of acoustic-based human behavior sensing and recognition systems is on the basis of the Doppler effect of the signal reflected by users [23–26]. Such systems do not have tracking capability and can only recognize predefined behaviors, as Doppler shift can only provide coarse-grained measurements of the speed or direction of user movements due to the limited frequency measurement precision.

The second line of acoustic-based systems is on the basis of decoding the echo of FMCW sound wave to measure the human body. Specifically, FMCW [27] indirectly estimates the propagation delay based on the frequency shift of the chirp signal, based on which we can measure the movement displacement of the target user, and a number of behaviors can be recognized [28]. However, the distance estimation resolution of FMCW is restricted by the sweep bandwidth; it is difficult to achieve very high distance estimation resolution (e.g., detecting human respiration with chest movement displacement of less than 1 cm) with narrowband commodity

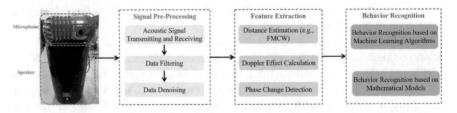

Fig. 4.3 A general framework of acoustic-based behavior recognition approaches [22]

acoustic device. To tackle this issue, a correlation-based FMCW (C-FMCW) approach [22] is proposed to achieve more accurate distance estimation. Unlike traditional FMCW, C-FMCW estimates the round-trip propagation time of acoustic signals by discovering the maximum correlation between transmitted signal and received signal. For digital FMCW signal, the round-trip propagation time can be measured by detecting the number of samples corresponding to the maximum correlation. As such, the distance estimation resolution of C-FMCW is limited only by the acoustic signal sampling rate. With current prevailing audio systems of 48 kHz, C-FMCW can achieve a ranging resolution of around 0.4 cm. The ranging resolution can be further improved if higher sampling rates (e.g., 96 kHz) are supported by the acoustic systems.

The third line of research uses OFDM pulses or CW signals to detect phase changes and facilitate real-time behavior tracking [29, 30]. For instance, by leveraging the OFDM pulses, the FingerIO system [29] achieves a finger location tracking accuracy of 8 mm and also allows 2D drawing in the air using COTS mobile devices. Similarly, the LLAP system [30] uses CW signals, which is less noisy due to the narrower bandwidth compared with OFDM pulses, allowing it to achieve better tracking accuracy.

In general, compared with other kinds of wireless signals (e.g., RF signal, optical signal), sensing human behavior using acoustic signals has the following advantages. (1) Short wavelength means more accurate behavior sensing. Due to relatively slow propagation velocity, even with 20 kHz transmitting frequency, the wavelength of transmitting acoustic signal is only about 1.7 cm. (2) It is easy to control the process of acoustic transmitting and receiving. For example, using commodity acoustic devices, we can directly control acoustic transmitting and receiving process by calling system audio interface. (3) Compared with the raw data (e.g., RSSI or CSI) of RF-based sensing systems, the data that we can obtain from speaker and microphone is the original transmitting and receiving signal, which contains more abundant information.

The disadvantages of using acoustic signal to sense and recognize human behavior can be summarized as follows. (1) Acoustic signal is one kind of mechanical wave, which attenuates severely when propagating in the air. This characteristic limits its sensing range. (2) Compared with RF signals, the acoustic signal has poor penetration ability.

From the above discussion, we know that acoustic is suitable for sensing fine-grained behavior in short range, such as human vital signal sensing, gesture recognition, and hand tracking.

References

1. Nickel, C., Busch, C., Rangarajan, S., & Möbius, M. (2011). Using Hidden Markov Models for accelerometer-based biometric gait recognition. IEEE, International Colloquium on Signal Processing and ITS Applications, pp 58–63.

2. Zanca, G., Zorzi, F., Zanella, A., & Zorzi, M. (2008). Experimental comparison of RSSI-based localization algorithms for indoor wireless sensor networks. The Workshop on Real-World Wireless Sensor Networks, pp 1–5.
3. Huang, Y. F., Yao, T. Y., & Yang, H. J. (2015). Performance of Hand Gesture Recognition Based on Received Signal Strength with Weighting Signaling in Wireless Communications. International Conference on Network-Based Information Systems, pp 596–600.
4. Wang, G., Zou, Y., Zhou, Z., Wu, K., & Ni, L. M. (2014). We can hear you with Wi-Fi!. International Conference on Mobile Computing and NETWORKING, pp 593–604.
5. Xi, W., Zhao, J., Li, X. Y., & Zhao, K. (2014). Electronic frog eye: Counting crowd using WiFi. INFOCOM, 2014 Proceedings IEEE, pp 361–369.
6. Z. Wang, B. Guo, Z. Yu and X. Zhou, "Wi-Fi CSI-Based Behavior Recognition: From Signals and Actions to Activities," in IEEE Communications Magazine, vol. 56, no. 5, pp. 109-115, May 2018. DOI: https://doi.org/10.1109/MCOM.2018.1700144
7. X. Liu, J. Cao, S. Tang, J. Wen and P. Guo. Contactless Respiration Monitoring via Off-the-shelf WiFi Devices. IEEE Transactions on Mobile Computing, 15(10): 2466-2479, 2016.
8. J. Liu, Y. Wang, Y. Chen, J. Yang, X. Chen, and J. Cheng. Tracking Vital Signs during Sleep Leveraging Off-the-shelf WiFi. ACM MobiHoc 2015, 267–276.
9. H. Li, W. Yang, J. Wang, Y. Xu, L. Huang. WiFinger: Talk to Your Smart Devices with Finger-grained Gesture. ACM UbiComp 2016, 250–261.
10. G. Wang, Y. Zou, Z. Zhou, K. Wu, and L. M. Ni. We Can Hear You with Wi-Fi. ACM MobiCom 2014, 593–604.
11. K. Ali, X. Liu, W. Wang, and M. Shahzad. Keystroke Recognition Using WiFi Signals. ACM MobiCom 2015, 90–102.
12. Y. Wang, J. Liu, Y. Chen, M. Gruteser, J. Yang, and H. Liu. E-eyes: device-free location-oriented activity identification using fine-grained Wi-Fi signatures: ACM MobiCom 2014, 617–628.
13. W. Wang, X. Liu, and M. Shahzad. Gait Recognition Using WiFi Signals. ACM UbiComp 2016, 363–373.
14. Y. Zeng, P. Patha, and P. Mohapatra. WiWho: WiFi-based Person Identification in Smart Spaces. IEEE IPSN 2016, 1–12.
15. H. Wang, D. Zhang, Y. Wang, and J. Ma. RT-Fall: A Real-time and Contactless Fall Detection System with Commodity WiFi Devices. IEEE Transactions on Mobile Computing, 16(2): 511-526, 2017.
16. J. Zhang, B. Wei, W. Hu, and S. Kanhere. WiFi-ID: Human Identification using WiFi signal. IEEE DCOSS 2016, 75–82.
17. T. Xin, B. Guo, Z. Wang, M. Li, Z. Yu, and X. Zhou. FreeSense: Indoor Human Identification with Wi-Fi Signals. IEEE GlobeCom 2016, pp. 1–6.
18. L. Sun, S. Sen, D.S Koutsonikolas, and K. Kim. WiDraw: Enabling Hands-free Drawing in the Air on Commodity WiFi Devices. ACM MobiCom 2015, 77–89.
19. W. Wang, X. Liu, M. Shahzad, et al. Understanding and modeling of Wi-Fi signal based human activity recognition. ACM MobiCom 2015, 65–76.
20. H. Wang, D. Zhang, J. Ma, Y. Wang, Y. Wang, D. Wu, T. Gu, and B. Xie. Human Respiration Detection with Commodity WiFi Devices: Do User Location and Body Orientation Matter. ACM UbiComp 2016, 363–373.
21. D. Zhang, H. Wang, and D. Wu. Toward Centimeter-Scale Human Activity Sensing with Wi-Fi Signals. IEEE Computer, 50(1): 48-57, January, 2017.
22. Tianben Wang, Daqing Zhang, Yuanqing Zheng, Tao Gu, Xingshe Zhou, Bernadette Dorizzi. C-FMCW Based Contactless Respiration Detection Using Acoustic Signal. ACM UbiComp 2018.
23. Sidhant Gupta, Daniel Morris, Shwetak Patel, and Desney Tan. Sound wave: using the Doppler effect to sense gestures. In Proc. ACM CHI, 2012.
24. Md Tanvir Islam Aumi, Sidhant Gupta, Mayank Goel, Eric Larson, and Shwetak Patel. Doplink: Using the Doppler effect for multi-device interaction. In Proc. ACM UbiComp, 2013.

25. Ke-Yu Chen, Daniel Ashbrook, Mayank Goel, Sung-Hyuck Lee, and Shwetak Patel. Airlink: sharing files between multiple devices using in-air gestures. In Proc. ACM UbiComp, 2014.
26. Tianben Wang, Daqing Zhang, Leye Wang, Xin Qi, Bernadette Dorizzi, Xingshe Zhou. Contactless Respiration Monitoring using Acoustic Signal with Off-the-shelf Audio Devices. IEEE Internet of Things Journal, DOI:https://doi.org/10.1109/JIOT.2018.2877607.
27. A. G. Stove. (1992). Linear FMCW radar techniques. Radar & Signal Processing IEEE Proceedings F , 139, 5:343-350.
28. Rajalakshmi Nandakumar, Shyamnath Gollakota, Nathaniel Watson. Contactless Sleep Apnea Detection on Smartphones. The 13th ACM Annual International Conference on Mobile Systems, Applications, and Services, 2015, 45–57.
29. Rajalakshmi Nandakumar, Vikram Iyer, Desney Tan, and Shyamnath Gollakota. FingerIO: Using active sonar for fine-grained finger tracking. In Proc. ACM CHI, 2016.
30. Wei Wang, Alex X. Liu, and Ke Sun. Device-Free Gesture Tracking Using Acoustic Signals. In Proc. ACM MobiCom, 2016.

Chapter 5
Individual Behavior Recognition

Abstract In this chapter, we present some of our recent research advances on individual behavior sensing and recognition. Specifically, in Sections 5.1 and 5.2, we present two human mobility-related works (i.e., mobility prediction and disorientation detection) by leveraging GPS trajectories. Afterwards, we discuss how to recognize human behaviors by using smartphones in Sections 5.3 and 5.4 (i.e., human-computer operation recognition and human localization), followed by two device-free sensing-based behavior analysis practices in Sections 5.5 and 5.6 (i.e., human identity recognition and respiration detection).

5.1 Human Mobility Prediction by Exploring History Trajectories[1]

5.1.1 Introduction

With the rising popularity of mobile sensors and portable devices, "serendipitous" social interaction is emerging and becoming pervasive, where interaction is triggered when two devices (i.e., two users) are located closely in a mobile peer-to-peer environment [1], or wireless ATM networks [2]. Studies have been conducted to support unexpected interaction utilizing users' mobile wireless devices (e.g., mobile phone, vehicle PDA, and wearable devices). In these studies, human interaction is facilitated by means of peer-to-peer data sharing between matching devices [3] (i.e., users), and the "matching" is defined by similarity computed from user profiles (e.g., music, pictures, and tagged preferences) on the phone. However, these mechanisms merely provide services for people who have common interests and thus cannot enhance human social relations in the physical world.

---

[1]Part of this section is based on a previous work: Z. Yu, H. Wang, B. Guo, T. Gu and T. Mei, "Supporting Serendipitous Social Interaction Using Human Mobility Prediction," in IEEE Transactions on Human-Machine Systems, vol. 45, no. 6, pp. 811–818, Dec. 2015. DOI: https://doi.org/10.1109/THMS.2015.2451515

The term "serendipity" was coined by the novelist Horace Walpole in the eighteenth century to describe unexpected and fortunate discoveries; "serendipity" was originally used to refer to making accidental discoveries when looking for one thing and finding another [4]. We, therefore, characterize the serendipitous social interaction as unplanned, not affecting users' schedules, bringing convenience, and creating positive emotion (e.g., happiness) in serving as an activity user, and users can choose to participate or not. For example, take two college students who are friends and who have not met for weeks. Both of them often visit the same library at the same time, but they do not recognize that. In such a scenario, if a mobile application helps them to capture and learn the serendipitous opportunities to meet and have a chat, the friendship between these two students will be enhanced.

To support such social interactions, we need first to discover the serendipitous interaction opportunities [5], similar to the communication channel of mobile intermediate nodes in an opportunistic network [6]. With the captured unplanned and transitory opportunities, serendipitous interaction can be used to make our lives easier. For example, if a user is aware that his/her roommate is passing by a grocery, he/she may ask him/her to buy something for him/her. However, this kind of serendipitous opportunity will disappear in a matter of seconds as he/she walks away. In this scenario, prediction, rather than instant behavior detection or notification, would be better. Mobility prediction typically leverages human trajectory data (e.g., GPS data, check-in records, and intercell [7]). The majority of mobility prediction methods focus on predicting a user's future mobility status (e.g., where and when a user arrives at the next venue, and duration). In this chapter, we predict users' future temporal and spatial contexts such as venue, arrival time, and user encounter. As user mobility status (e.g., location and time) may change in a very short period of time in the physical world, coarse-grained (e.g., hour-level temporal prediction error) and low-accuracy (e.g., accuracy of next venue judgment) mobility prediction cannot satisfy the needs of practical applications. Aiming to achieve high accuracy and low error, we discover the strong spatial and temporal regularities in GPS trajectories and use supervised learning algorithms to train historical mobility instances (generated by crowd users).

Few applications have been proposed, especially any combining spatial and temporal prediction information. In this chapter, we aim to facilitate serendipitous social interactions and the real participation activities in the physical world, which is different from simple information sharing mechanisms in existing traditional applications. Leveraging the overlap and regularity in collected GPS trajectories, we deploy supervised learning algorithms and achieve an accuracy of over 90% for predicting a user's next venue, and minute-level (i.e., an average of about 5 min) prediction error for arrival time.

By leveraging mobility prediction to forecast the occurrence of serendipitous interaction opportunities, we propose a three-layer framework to support serendipitous social interaction, develop two applications for use on a university campus, and conduct a survey about the application. The main contributions of this chapter are summarized as follows.

- We propose a system framework for supporting social interaction by means of facilitating users to participate in interaction activities in the physical world. Under this framework, mobility prediction is introduced to capture serendipitous interaction opportunities.
- We leverage mobility prediction to discover serendipitous social interaction opportunities. Based on the spatiotemporal mobility regularities in users' trajectories, we first predict where and when a user will arrive, and then, we can determine if some users may soon encounter (i.e., have the same destination).
- We develop the prototype of the proposed framework and build two applications based on the framework, HelpBuy and EaTogether, to support serendipitous social interaction on campus.

5.1.2 Related Work

Previous studies attempted to facilitate online social interactions in social networks. Under the premise that users who have the same interests are more willing to interact with each other (i.e., user homogeneity), studies try to help users find interaction with friends through personal preference similarity calculation in social networks, in which social interactions are facilitated in the form of learning [8], date [1, 9], and travel [10]. These kinds of applications only enhance social interaction within an online social network, rather than establish social connectivity in the physical world.

On the other hand, researchers focus on supporting face-to-face human interaction, in which social interactions are facilitated by leveraging serendipitous communication opportunities between mobile devices (i.e., users). BlueFriend [11] is an application that leverages Bluetooth to find friends among nearby users. Bluedating [9] provides localized dating services to help users find desired partners. Paradiso et al. [12] developed a badge system, which is equipped with wireless infrared and radio frequency networking, to facilitate social interaction between wearers. Lawrence et al. [3] developed three applications by exploiting the "co-presence" interactions (i.e., incidental interactions) between mobile devices. In the mobile peer-to-peer environment, as Yang et al. [10] proposed, information sharing and social interaction are facilitated by capturing serendipitous interaction opportunities. However, interaction opportunities are always discovered by device detection. Devices are the actual participant in these applications, rather than the humans themselves. In this chapter, we predict serendipitous interaction opportunities leveraging mobility prediction using users' current mobility status and present mobile applications that facilitate people actively participating and interacting in the physical world. Mobility prediction is the main supporting technology of our system. It aims to discover serendipitous interaction opportunities.

Mobility prediction has been studied to perceive human future mobility status (e.g., next venue, arrival time, and duration) from different perspectives. Xiong et al. [13] took advantage of the similarity between people's trajectories and proposed collective behavioral patterns for improving prediction accuracy. Do and Gatica-

Perez [14] adopted a factorized conditional model according to the extracted features, which can reduce the size of the parameter space as compared to conditional model. Cho et al. [15] discussed the contribution of a location-based social network and an individual user's periodic movement pattern in mobility prediction and developed a model that combines the periodic day-to-day movement patterns with the social movement effects coming from the friendship network. Noulas et al. [16] used check-in data in Foursquare and proposed a set of mobility prediction features to capture the factor that drives users' movement, and achieved around 90% prediction accuracy. Baumann et al. [17] analyzed the influence of temporal and spatial features in mobility prediction and predicted transitions. McInerney et al. [18] presented a Bayesian model of population mobility to tackle the data sparsity problem in mobility prediction. Song et al. [19] utilized the Order-2 Markov predictor with fallback and obtained a median accuracy of about 72% for users with long trace lengths. Song et al. [19] provided evidence that the prediction accuracy of an individual's next location had an upper bound of 93%. Lin et al. [20] reported on a study of GPS data-based mobility prediction accuracy and suggested that a predictability upper bound of 90% is able to support ubiquitous applications. Considering the sparsity of trajectory data, such as Call Detail Records [21] and taxi GPS data, researchers also made much progress. Wang et al. [22] proposed a mobility gradient decent approach and predicted user's destination, by using sparse taxi trajectories. Recently, some works explore mobility prediction using deep learning [23, 24]. By discovering region of interest (ROI) in the city, Jiang et al. [24] proposed a deep ROI-based modeling approach for effectively predicting human mobility. In this chapter, we predict not only an individual user's next venue, but also the arrival time and multiple users' encounters based on users' current mobility features (CMFs). We achieved a high prediction accuracy of around 90%, which enables discovering opportunities to support serendipitous live interactions.

5.1.3   Serendipitous Social Interactions Supporting System

5.1.3.1   Framework

The framework consists of three layers, as shown in Fig. 5.1.

Data Layer: In the data layer, a GPS dataset of trajectories is collected from users by the GPS data logger. Each GPS point has several spatial and temporal attributes, such as date, time, latitude, longitude, and speed, and trajectories with thousands of GPS points are recorded. These mobility data reflect the regularity of users' daily visits (i.e., the GPS point sequences in some venues) and the movement (i.e., the GPS point sequences between two venues). After data denoising, historical mobility instances are extracted from the GPS dataset for mobility prediction.

Mobility Prediction Layer: Repeatable behaviors and patterns exist in the GPS trajectories. For example, at a campus crossroad during the morning, if a student is

Fig. 5.1 System framework

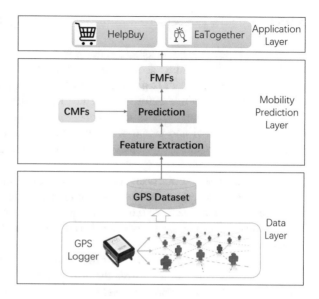

walking westward, then we predict that he/she will go to the gym; if he/she is walking towards the east, then we predict that he/she will arrive at the library. Moreover, people living in the same environment (e.g., students in a dormitory) may have the same regularity, and thus, users' mobility trajectories can be utilized as historical instances to predict the next venue and arrival time.

Human GPS trajectory contains many temporal (i.e., morning) and spatial (i.e., at a crossroad) context features. It is important to select the appropriate features to achieve high prediction accuracy and low complexity. As trajectories have regularity, we apply learning algorithms to discover patterns. Then, after perceiving individual users' CMFs (e.g., time, latitude, and longitude), the inference models can predict users' future mobility features (FMFs) (e.g., next venue and arrival time). At the individual level, we can predict a user's future status. Then, at the crowd level, we can predict multiple users' occasional encounters at a next venue (e.g., two users will meet at a restaurant in 10 min), which is similar to predicting data source location in [25].

Application Layer: In this layer, we design and develop applications to facilitate serendipitous interactions based on the prediction results of user's FMFs from the mobility prediction layer. For example, we can develop more efficient participatory sensing [26] applications with perceiving users' future positions. Furthermore, we can discover unexpected opportunities with user encounter prediction results. In this chapter, we implement two applications that facilitate social interactions in specific scenarios on a campus.

5.1.3.2 Mobility Prediction

We observe repeatable behavior and patterns in the GPS trajectories. For instance, a person is going to the same next venue when he/she is at the same position at the same time on different days (e.g., at noon, students at the same place are likely to go to restaurant; in the 5 min before class, students on different paths are all moving towards the classrooms). These real phenomena imply that students' trajectories contain strong spatial and temporal regularities. We explore these regularities for mobility prediction and apply supervised learning algorithms to train the inference model.

In this chapter, we predict users' FMFs at two granularities: individual and crowd. At the individual level, we predict where a user is going to (i.e., next venue) and when the user will arrive there (i.e., arrival time). At the crowd level, we predict the intersection of multiple users' trajectories (i.e., user encounter) based on the individual user mobility prediction.

Figure 5.2 shows the detailed mobility prediction procedure. First, historical mobility instances are extracted from the historical GPS dataset to form the training dataset. Second, since supporting serendipitous social interaction requires high prediction accuracy both spatially and temporally, we adopt learning algorithms to train using the historical dataset. The prediction task contains discrete (next venue) and continuous (arrival time) output; therefore, we use different learning algorithms to train inference models to predict each FMF. According to ref. 17, more CMFs are not necessary for higher prediction accuracy. We find that we need to select the suitable CMF set to achieve high accuracy and low complexity. Experiments are conducted by selecting different CMF sets from Table 5.1 and comparing their performance. Finally, after predicting results of the user's FMFs (i.e., next venue and arrival time) with the inference models, we can predict user encounter at the next venue.

Taking every GPS point in the trajectories as a mobility instance, we extract fine-grained attributes (e.g., latitude, longitude, time, direction, and speed of a GPS point) as the mobility instance's features. Leveraging the overlap and regularity in these trajectories, the mobility instances' features are highly cohesive and exhibit low coupling.

Essentially, each mobility point is a GPS point recorded on the path between two venues with a timestamp and GPS coordinate. However, in this chapter, a mobility point in fact has eight features (see Table 5.2). The former five features are all CMFs, which are originally recorded by GPS sensors, and the latter three features are all FMFs, @NV, @AT, and @UE, which are acquired. Specifically, the historical mobility points' FMF values are assigned using feature extraction steps to form training instances, while the mobile phone users' FMF values are to be predicted through inference models in practical applications. The CMFs are mutually independent.

None of the mobility features are related to personal information as limited by the data scale of individual users. Inspired by the collective behavior pattern [13] and the population modeling mechanism [25], we are able to achieve high prediction accuracy to meet the demands of application scenarios.

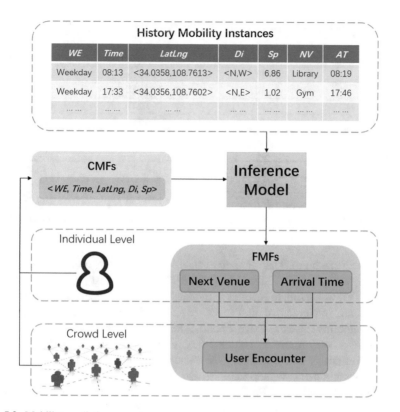

Fig. 5.2 Mobility prediction procedure

Table 5.1 Mobility features

Feature	Description	Example
@WE	Weekday or weekend	Weekday/weekend
@Time	Current time	18:06:08
@LatLng	Latitude and longitude	<34.0356, 108.7600>
@Di (direction)	N(north)/S(south), E(east)/W(west)	<N, E>
@Sp (speed)	Discretized into 4 levels, separated by 2, 5 and 10 (km/h)	Stroll(0–2)/walk(2–5)/scurry (5–10)/trot(>10)
@NV	Next venue	Library
@AT	Arrival time of next venue	18:14:34
@UE	Who, where, and when will meet	<Tom, restaurant, 11:56:32>

Predicting Next Venue: People's daily life generates almost the same life track every day, and their GPS trajectories exhibit high spatial and temporal overlapping and regularities. Specifically, a user's next venue is predominantly determined by his/her current spatial and temporal context (e.g., time and location), and people are likely to make the same decision when they are in the same situation (e.g., same

Table 5.2 Training decision tree on different CMF sets (f1: time, f2: LatLng, f3: direction, f4: speed, f5: WE)

Feature sets	Leaf number	Size of tree	Accuracy (%)
{f1}	497	993	35.46
{f2}	922	1843	48.59
{f1, f2}	4621	9241	89.73
{f1, f2, f3}	5405	10,547	84.82
{f1, f2, f4}	5198	9849	88.64
{f1, f2, f5}	4388	8775	**90.62**
{f1, f2, f3, f5}	5337	10,415	85.99
{f1, f2, f4, f5}	4940	9397	89.64
{f1, f2, f3, f4, f5}	5775	10,730	84.47

The most effective CMF feature sets are {f1, f2, f3}, based on which the best performance of next venue prediction is obtained

location and same time). Therefore, to evaluate between users' mobility features and next venue, we choose Decision Tree [27, 28], Random Forest [29], KNN [29], and BayesNet [30] as candidate prediction models. These four supervised classification algorithms are implemented in WEKA [31]. By deploying these algorithms on historical mobility instances, we obtain inference models (see Fig. 5.2). These models take a user's mobility contexts (i.e., CMFs, e.g., time, latitude, longitude) as input conditions and determine the venue as output. We use tenfold cross-validation correct percentage to measure accuracy of predicting the next venue.

Predicting Arrival Time: Predicting arrival time accurately is crucial. For example, predicting users' future location and arrival time provides location-based services to have a sufficient start-up time. Similar to the next venue prediction procedure, arrival time is also related to the user's mobility context; therefore, we need supervised classification algorithms to determine the arrival time. As the output of arrival time is a continuous value, we employ linear regression [29] and model tree [18, 32] in WEKA [31], to predict the arrival time. We use tenfold cross-validation to compare their performance in terms of arrival time prediction error.

Predicting User Encounter: If the users' mobility features, such as next venue and arrival time, can be predicted with high accuracy, then we predict people's encounters. We define an encounter as follows:

$$\text{Encounter}(s_1, s_2) \Leftrightarrow NV_1 = NV_2 \&\& |AT_1 - AT_2| < \text{To}, \qquad (5.1)$$

where s_1 and s_2 are two users, NV_1 and NV_2 are the predicted next venues of the two users, AT_1 and AT_2 are the predicted arrival time, and To is the preset time threshold that the former arriving user would wait for the latter one. Combined with friend relationship information (i.e., friendship network as shown in Fig. 5.2), the encounter prediction results can be utilized in applications to support serendipitous social interactions between friends, unplanned in advance.

Overall, serendipitous social interaction opportunities (e.g., people encounters) are discovered from mobility prediction results and utilized in specific applications in the upper layer to support and enhance human interaction.

5.1.4 Application

We develop the prototype of the proposed framework. In this section, we present two prototype applications built on the framework to support serendipitous interactions on a campus. *HelpBuy* is a mutual application whereby a user requests others to buy what he/she needs, while others are incidentally on their way nearby a point of sale. *EaTogether* provides reminder services for users to find unplanned chance to eat together with their friends.

5.1.4.1 HelpBuy

HelpBuy is an application taking advantage of the next venue prediction. A user can request another to buy something along their way.

Usage scenario. One morning, Henry is working hard in the school laboratory. He wants a cup of milk, but he prefers not to leave the lab to go to the supermarket, which is far away. First, as shown in Fig. 5.3a, Henry points out where to buy the milk on the map (BuyPlace, Δ) and where the milk should be delivered to (Destination, o, i.e., Henry's current position, the lab building). He also needs to provide a description of what to buy (i.e., milk, 500 mL) and click the "Search" button. Then, as shown in Fig. 5.3b, the users who are willing to help buy (i.e., volunteers, $\star$, near the BuyPlace and predicted to arrive at the Destination) appear on the map. Finally, one of the volunteers accepts Henry's request and fulfills the *HelpBuy* task.

In this application, people's willingness to share their real-time mobility status is critical. In order to encourage people (acting as the role of volunteer) to help others (acting as the role of clients), an incentive mechanism is needed. The client offers a reward when he/she publishes a task, and the volunteer earns reward points by fulfilling tasks.

5.1.4.2 EaTogether

EaTogether is an application that predicts when friends can have encounters at restaurants and recommends them to have a meal together.

Usage scenario. Three students, Athos, Aramis, and Porthos, are friends. They study on the same campus but they have not met together for a long time. The system discovers if there is a chance for them to meet and have a chat with each other, for example, occasionally meeting at a campus restaurant and having lunch together. In the past, they had their lunches at the same restaurant at the same time, but were not aware. Now, with *EaTogether*, they can capture every such opportunity to meet and have a meal.

One day at noon, *EaTogether* predicts that Athos is on the way to the restaurant and, then, asks Athos whether he wants to share his mobility status and try to discover unplanned chances for having lunch together with his friends as shown in

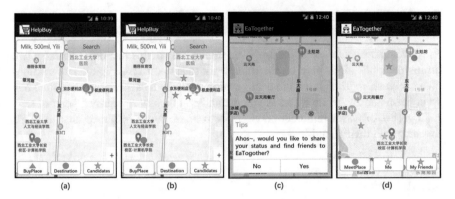

Fig. 5.3 Two applications. (**a**) HelpBuy (before search), (**b**) HelpBuy (after search), (**c**) EaTogether (tips), (**d**) EaTogether (encounter notice)

Fig. 5.3c. With Athos' permission, the backend server predicts that his friends, Aramis and Porthos, are heading towards the same destination, and then refreshes the screen of Athos's phone with an encounter notice, as shown in Fig. 5.3d, in which *MeetPlace*, ∗∗, means the venue where they will meet (i.e., their same *next venue*), yellow ★ denotes the user himself (i.e., Athos), while orange ★ denotes his friends. In this application, we also show when the user and his friends will arrive at the venue.

Apart from next venue and arrival time prediction results, the users' relationship information is needed in this application. We can collect users' friendship information from online social networks, e.g., Facebook. By facilitating friends' unplanned interaction in this way, *EaTogether* is able to enhance, or strengthen, their friendship.

5.1.5 Performance Evaluation

5.1.5.1 Data Collection

We collected GPS trajectory data generated by student volunteers in a 1000×1000 m^2 campus, where 17 popular venues are selected as the next venue candidates. We classified these venues into four categories: restaurant, dormitory, workplace, and sports. Each volunteer carried a portable GPS logging device for several days, and the device recorded their daily trajectories.

The obtained dataset includes trajectories from 156 volunteers and consists of about 2.8 million GPS points. These GPS points are converted to 64,482 mobility instances after preprocessing, which are used to train inference models. Utilizing these models, we can predict FMFs based on a user's CMFs (e.g., time, latitude, and longitude).

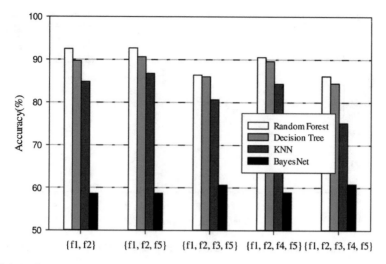

Fig. 5.4   Performance of different models (f1: time, f2: LatLng, f3: direction, f4: speed, f5: WE)

5.1.5.2   Results of Next Venue Prediction

We adopted tenfold cross-validation to evaluate the performance of different prediction models, as shown in Fig. 5.4, which demonstrates the result under five different CMF sets. Since the lifestyle of students is relatively fixed, the collected GPS trajectories contain strong temporal and spatial regularities. Therefore, we achieved good prediction performance by using supervised classification algorithms. In general, random forests, decision trees, and KNN yield similar performance, while BayesNet performs poorly since the features are mutually independent. Considering the complexity of different models, we finally choose decision trees for venue prediction.

Furthermore, to achieve high prediction accuracy and low complexity, we need to compare the performance of different feature sets. Table 5.2 illustrates the training results of different CMF sets. The leftmost column shows the feature set, the middle two columns are from the decision tree, and the rightmost column indicates the tenfold cross-validation accuracy. Accordingly, we find that time and location (i.e., LatLng) are key features for next venue prediction, confirming the importance of temporal and spatial context in mobility prediction. We also find that direction and speed have negative effects, which might be due to GPS signal noise. Meanwhile, since daily schedules of college students are different on weekdays and weekends, the accuracy improved by 1% when WE is considered. As a result, the feature set {f1, f2, f5} is selected as the CMFs for next venue prediction.

We randomly selected 2000 mobility points as test instances to measure the prediction accuracy of the developed model, and the overall accuracy was 92.1%. Specifically, the prediction accuracies for different kinds of venues were 90.6% (restaurant), 88.1% (dormitory), 94.7% (workplace), and 91.5% (sports), respectively. In general, the next venue prediction accuracy is higher than 90%.

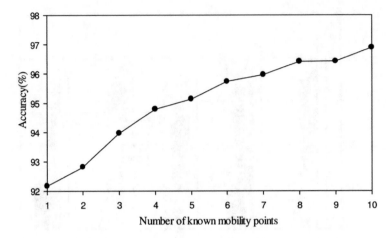

Fig. 5.5 Next venue prediction accuracy with different numbers of known mobility points

Figure 5.5 shows the accuracy of next venue prediction with different numbers of known mobility points. If we know more about a user's current mobility status, we can predict one's next venue more accurately. Nguyen et al. [33] also adopted different classification algorithms for next location prediction. Their reported accuracies were 69.0% (decision tree, C4.5/J48) and 54.1% (KNN, where K is empirically set to 5), which were much lower than our results. The reason is that our GPS dataset has higher regularity and contains a set of fine-grained spatial and temporal mobility features. Specifically, our results also comply with the suggested predictability upper bound, i.e., 90% [20].

We have also evaluated the performance of arrival time prediction, user encounter prediction, as well as two applications (*HelpBuy* and *EaTogether*); please refer to the article [34] for more details.

5.2 Disorientation Detection by Mining GPS Trajectories[2]

5.2.1 Introduction

With the rapidly aging population in most countries, the proportion of elders (i.e., people aged 65 and above) in the world will significantly increase in the next few decades [35]. As people grow older, a considerable number of elders would experience physical and cognitive impairments. Dementia has been identified as a progressive, disabling, and chronic disease, affecting 5% of elders aged above

---

[2]Part of this section is based on a previous work: Q. Lin, D. Zhang, K. Connelly, H. Ni, Z. Yu, X. Zhou, Disorientation detection by mining GPS trajectories for cognitively-impaired elders, Pervasive and Mobile Computing, 19, 2015, Pages 71–85

65 and around 40% of elders aged over 90 [36]. Generally, an elder with dementia performs her daily routines poorly, and often cannot conduct instrumental activities of daily living [37], such as preparing meals or doing housework.

One of the most challenging problems faced by elders with dementia is navigating or remembering landmarks, making it a primary difficulty when doing daily outdoor activities [38], such as visiting doctors and going shopping. Disorientation, or getting lost, occurs frequently in unfamiliar locations for people with mild dementia; for those with moderate cognitive impairment, it may even happen in familiar environments [39]. As reported by McShane et al. [40] in their longitudinal study on 104 participants over 5 years, "getting lost" has a prevalence rate of 24% in the first year of the onset of dementia. As a result, safety has become the critical concern for cognitively impaired elders, particularly when they conduct routine activities in outdoor environments. Frequent cases of missing elders have indicated that the traditional solutions to the disorientation problem, such as wearing an identity card [41], do not work very well.

Continuously tracking elders' activities and delivering personalized services (e.g., prompting or calling for help under an emergency circumstance) using pervasive technologies provide a promising solution to many elderly care problems. Personalized services are able to improve the autonomy and independence of elders while reducing the potential risks and carers' workload. Some work in this area has been reported in the literature, including preventing elders from walking out of a "security" area by positioning RF-based detectors on exits [42] or letting elders carry a body-attached sensor such as GPS (Global Positioning System)-equipped artifact [43]. Another line of research is to provide outdoor navigation support for elders when deviating from pre-defined routes by using a location-sensing device (e.g., GPS-equipped cell phones) combined with GIS (geographic information system) [44, 45].

While limiting the freedom of movement to prevent elders from being lost is impractical in real life for those with mild dementia who still care for themselves most of the time, the solutions in [44, 45] can only detect part of elders' disorientation behaviors. In this work, we propose a method that is able to find potential disorientation behaviors by monitoring an individual's movement trajectory. This detection method ensures that the disorientation behaviors (i.e., getting lost) can be detected in real time, assuming that elders are equipped with GPS-enabled mobile phones. Our work is based on the following observations.

- Despite differences, humans follow simple patterns in their daily routines [46]. This is particularly true for elderly people with dementia because most of them maintain a simple lifestyle. This fact makes it possible to extract hidden spatiotemporal patterns from the recordings of daily routines and model one's mobility pattern.
- The wide coverage of infrastructure (e.g., GPS satellites) and high penetration of personal mobile computing platforms (e.g., cell phones or PDAs) make it possible to collect human movement information in real time. Thus, appropriate and timely alerting services can be provided when elders are disorientated.

As human mobility is highly random and different people have different habits with respect to the places they visit and the routes they take, it is a very challenging task to characterize people's mobility behaviors with one unified and rigorous model. The second challenge is to characterize the common features of disorientation, examining the huge difference among individual's normal and disorientation behaviors. The third challenge is to develop a simple but effective algorithm which is able to tell the disoriented trajectory from the normal ones in real time, so that timely alerting services can be provided to elders.

To address the above challenges, in this chapter we project each elder's historical movement trajectories onto grid cells in a digital map, and model each elder's regular movement as a graph where vertices represent frequented places and edges correspond to the routes between places. In this way, each individual's movement trajectory can be represented as a sequence of traversed cell symbols and disorientation detection becomes a problem of comparing the ongoing movement trajectory against each user's regular movement graph. Second, we analyze various user disorientation behaviors and discover that all elders' disorientation trajectories contain outlier routes either deviating from the model, or looping inside the graph, or a combination of the two. Finally, we propose an effective disorientation detection method based on the isolation-based mechanism [47] to recognize the ongoing outlier routes corresponding to disorientation behaviors. Specifically, our method evaluates an ongoing test trajectory through mining and monitoring its support degree in the existing historical trajectory dataset. Those trajectories with low support degree are regarded as outliers, i.e., a trajectory corresponding to a disorientation behavior. In summary, the main contributions of this work include the following:

- First, we propose to model the regular movement pattern of each elder as a graph and characterize the features of disorientation behaviors as containing outlier routes either deviating from the model, or looping inside the graph, or a combination of the two.
- Second, we convert the disorientation detection problem into comparing the ongoing movement trajectory against the regular movement graph.
- Third, we propose an isolation-based detection method, which identifies disorientation by mining and monitoring its support degree in the existing historical trajectory dataset.
- Lastly, we use ten individuals' real-world GPS dataset to evaluate the proposed method. The experimental results demonstrate that our method performs well in terms of accuracy (i.e., the detection rate and false-positive rate), as well as time complexity.

5.2.2 Related Work

This section reviews related work in three relevant sub-areas: outdoor monitoring, route finding, and anomalous GPS trajectory detection. The research in the first category was aimed mainly at monitoring safety of elders with severe dementia in outdoor environments, whereas work in the second category is intended to provide navigation support for people with mild cognitive decline. The final category focuses on research in detecting anomalous trajectories.

5.2.2.1 Outdoor Monitoring

There has been much work to prevent people with severe dementia from getting lost by constraining them to a protected area. Generally, a location-sensing device (custom-built or off-the-shelf) is needed to collect a user's location information.

Shimizu et al. [48] developed a location system by utilizing a location-sensing device and GIS. The patient carries a responder consisting of a cell phone and GPS unit. Carers can find the patient's current position on a map displayed on their PC by invoking the responder through the cell phone network. This system was developed mainly to alleviate carers' workload when caring for patients with severe dementia. Ogawa et al. [43] investigated a safety support system by locating a wandering person's position via a mobile phone terminal (having no telephone call functionality). The terminal obtains a cell station ID and sends it to the phone company, allowing the PC to download the latitude and longitude data of the user's location (i.e., station location) from the mobile phone company via the Internet. When the wandering elderly person is away from home, the system automatically informs the caregiver and also sends the elder's location by e-mail to the carers. For further determining the user's surroundings, in their recent work, they added a small microphone to collect environmental sound [49]. In these systems, the authors report that the wandering person's location can be identified within 100 m from the mobile phone company's antenna ID.

There are also systems that attempt to assist older adults directly. For example, the Take Me Home-Service in the COGKNOW project [50] aims to help disoriented users to find their way back home by utilizing a GPS-equipped PDA. Similarly, researchers have investigated a mobile social network [51] that provides monitoring for elders using a GPS-enabled mobile phone. In this system, when an elder is beyond the pre-defined security perimeter around their home, their mobile will ring and vibrate, displaying a map on the mobile on how to get home.

The above solutions were developed primarily for people with severe dementia, either setting tight constraints on where a patient may be or providing a way of sharing their current location with a carer.

5.2.2.2 Route Finding

Finding a way typically requires a person knowing where he/she is, knowing his/her destination, following the best route to the destination, recognizing the destination upon arrival, and finding his/her way back [52]. All of these are challenging tasks for the cognitively impaired elders due to declines in their memory, thought, and reasoning functions.

The Opportunity Knock system proposed by Patterson et al. [44] is one of the earliest works in this field. In this system, cognitively impaired people are tracked in real time with a GPS device. Location data is sent to a remote server via a cell phone. The inference engine running on the server detects potentially erroneous behavior, such as taking an incorrect bus, based on user's historical trajectories. If an anomaly is detected, the system plays a sound like a knock on a door to get the user's attention, and then recommends a way to get back on track to the user. The user's likely transportation mode is learned from their historical trends and a hierarchical dynamic Bayesian network model [53]. Similarly, Chang [45] investigated detecting deviation from traveling trajectories for cognitively impaired workers. By comparing every incoming GPS point with a normal path, a trajectory with deviation can be detected in real time and reminders are provided to the user if needed. However, the chapter did not specify how to identify normal paths. Based on this work, a mobile location-based networking system was introduced by Chang and Wang [54] to further improve services for cognitively impaired workers by including formal caregivers in the system. Request for help can be automatic (i.e., when the system detects a potential deviation) or manual (i.e., when a person feels lost).

With both the model-based [44] and distance-based [45] approaches, there is a high false-positive rate due to the abundant noise of GPS data and the randomness of human mobility. In addition, model-based methods are usually not optimal in detecting anomalies [47], while distance-based methods suffer from a high-time complexity due to the frequent calculation of distances. Furthermore, both approaches are incapable of detecting a looping pattern that occurs often when people are disoriented.

5.2.2.3 Outlier Trajectory Detection

The work in this area focuses on detecting anomalous trajectories. Currently, there are a large number of solutions, each addressing different aspects of abnormality by targeting specific problems.

For instance, Lee et al. [55] proposed the partition-and-detect framework to detect outlier sub-trajectories. In their framework, both the distance and density measures are used for anomaly detection. Like Lee et al., Ge et al. [56] discovered evolving trajectory outliers based on an outlier score by combining the evolving direction and density of trajectories. Using a cluster join mechanism, Bu et al. [57] presented an outlier detection framework for monitoring anomalies over continuous trajectory streams. Alternatively, Li et al. [58] developed a temporal outlier detection approach for vehicle traffic data, which aims to discover anomalous traffic changes in road networks. In addition, a number of different machine learning-based approaches [59–61] have been developed, with the limitation of high costs in labeling. Focusing on taxi applications, Zhang et al. [62, 63] proposed a significantly different approach to detecting anomalous trajectories by exploiting an isolation-based mechanism [47]. After modeling a trajectory as a sequence of traversed cells, anomaly detection is converted to a problem of finding those trajectories that are "few and different" from normal trajectory clusters. Similarly, in another GPS-based taxi application, Yuan et al. [64] developed a smart driving direction system that can help taxi drivers choose driving directions by providing a fastest route to a given destination at a given departure time based on GPS traces and intelligence of experienced drivers. An extended discussion about semantic trajectory molding and analysis has recently been done by Parent et al. [65].

Our work is directly inspired by Zhang et al.'s approaches [62, 63], but with a different research objective and model. Our work models one's movement trajectories as a graph and the objective is to discover disorientation behaviors for cognitively impaired elders, resulting in different situations (e.g., two types of outliers with certain duration) that should be taken into account.

5.2.3 Disorientation Detection Problem Formulation

5.2.3.1 Modeling Human Mobility as a Graph

As mentioned above, the elderly often pursue a simple lifestyle in their daily life by visiting relatively fixed places where they perform daily activities. As such, we are able to construct a graph to represent an individual's movement behaviors as (5.2)

$$
\begin{cases}
\text{TG} = (V, E) \\
V = \{P_1, P_2, \ldots, P_n\}, \\
E = \{(P_i, P_j), 1 \leq i \neq j \leq n\}
\end{cases}
\tag{5.2}
$$

where each elder's historical trajectory graph TG represents his/her movement model, where the vertex set V denotes physical places frequented by the elder, and the edge set E indicates paths between vertices. Specifically, the physical place $P_i \in V$ ($1 \leq i \leq n$) refers to a location an elder frequents in his/her daily routine, including places like his/her home and his/her favorite restaurant; the edge $(P_i, P_j) \in E$ denotes the traversed path from one of the places in V to another. Because the user movement model TG is created by extracting all the frequented places and traversed paths, the vertices and edges included in TG reflect the movement traces of one's historical trips. Each elder has a different movement model and he/she visits the places and routes at different frequencies. In order to distinguish vertices from other places, we call those in V semantic places in this chapter.

Assume that H is the elder's home (as shown in Fig. 5.6); on a regular basis, she visits the *community center* (C) for social engagement, the *grocery store* (G) for food, and the *department store* (S) for shopping. Sometimes, she goes to the *medical clinic* (M) for a health check, and *daughter's home* (D) for a gathering. In this case, we obtain an instance $\{H, C, S, D, G, M\}$ for set V and $\{(H; C), (H; G), (H; S), (H; M) \ldots, (S; D)\}$ for set E. A full trip is therefore an ordered sequence of paths in E, forming a closed loop that starts at and finishes in the elder's own home H. Consequently, for example, a trip to M can be either $\{(H; M), (M; H)\}$ *or* $\{(H; M), (M; G), (G; H)\}$; it is completely dependent on the lifestyle as exhibited by one's mobility history. All the normal trips are those which follow frequently one of the historical trips.

For elders' movement behaviors, we need to explore all potential outliers that may relate to disorientation behavior directly. Firstly, a trip to a new place obviously did not match the defined model, such as the trip t_1 (from H to d_1) in Fig. 5.6, because d_1 and part of t_1 are not contained in model TG. Therefore, trips like t_1 are outlier (i.e., few and different) as compared to historical trips. This happens because of either incorrect paths or traveling to a newly visited place. The former is likely to correspond to a disorientation behavior, while the latter should be considered closely by the end-user application since cognitively impaired elders are prone to experience adverse events, especially in unfamiliar settings. For this reason, the rare event of traveling to a new place is flagged as an outlier.

A second potential outlier is a trip in which the elder frequently reverses direction within certain routes, such as t_3 on path $(M; D)$ or t_2 around S. This type of behavior could mean that the elder feels confused about where to go, so he/she changes direction in a route frequently. It is reasonable to classify trips like t_2 and t_3 into outlier clusters for an elder with dementia if these trips never appear in his/her mobility history. However, since direction change also occurs after visiting a specific location, our algorithm excludes changes at semantic places. Additionally, since it is common for someone to change direction when they realize they have forgotten something, we test if the number of direction change is larger than a certain threshold before it is considered as an outlier. For example, while the first two changes in trip t_3 may correspond to the fact that the elder changes his/her mind or forgets something, the third change is likely to mean that the elder has lost his/her way. If the direction changes occur exactly in the same places, the fourth change rather than the third change is considered as the sign of disorientation. It is noteworthy that the physical distance between two changes should be large enough to distinguish disorientation from pacing behavior [66].

A final potential outlier is the situation in which an elder takes several paths that would be considered normal on their own, but performed in a different order as compared to historical trips. For instance, for a trip I containing $\{(H; G), (G; D), (D;$

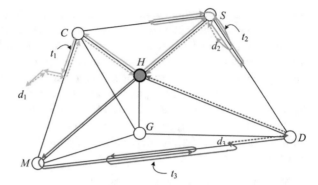

Fig. 5.6 A graph model of individual outdoor trips (H: home, C: community center, S: department store, G: grocery, M: medical clinic, and D: daughter's home)

M), (M; G), (G; D), (D; M)}, although any separate path in this trip is normal according to the defined model, it is still an outlier due to the incorrect order in which they appeared. This can happen when an elder becomes confused about what they are out to accomplish, so he/she keeps visiting familiar places, but with no clear goals. Trips falling into this category contain neither new places nor new direction changes as mentioned in the first two cases.

To sum up, an elder's disorientation-related trips are characterized by spatial deviation (e.g., strange places), repetitive changes of moving direction (e.g., getting lost between two places), as well as incorrect order of places visited (e.g., forgetting the purpose of their outing). The paths in these three categories are all corresponding to new traveling routes as compared to the historical trips.

5.2.3.2   Disorientation Behavior

Disorientation has different definitions in different application fields. As an example, the American Heritage® Medical Dictionary defines disorientation as "loss of one's sense of the direction, position, or relationship with one's surroundings." In this work, we are interested in detecting disorientation in elders' daily trips. Before giving a formal definition for elder's disorientation behavior, we classify the user movement behaviors into the following three categories:

- Regular: Moving from one semantic place to another following one of the frequent historical traces or paths
- Deviating: Moving to place along a route that has never been visited before as compared to historical traces
- Wandering: Traveling to semantic places or along old routes with a different sequence inside historical traces

In the above three patterns, the regular pattern refers to the normal movement behavior, while the remaining two patterns are viewed as disorientation behavior in this chapter. Specifically, a trip with a deviating pattern corresponds to either exploring a new place purposely or moving to an incorrect place with unfamiliar routes. For the wandering behavior, it includes three different movement patterns: (1) A trip contains direction changes inside a path between two semantic places (see t_3 in Fig. 5.6), where the direction change never happens in the historical traces; (2) a trip contains direction changes around a semantic place (see t_2 in Fig. 5.6), where the direction change never happens in the historical traces; and (3) a trip contains at least one old path and two semantic places but the trip never occurs in the historical traces (such as $I = \{(H; G), (G; D), (D; M), (M; G), (G; D), (D; M)\}$ in Fig. 5.6).

Based on these two types of disorientation patterns (i.e., deviating and wandering), we now formalize the definition of disorientation for cognitively impaired elders as follows:

Definition 1 (Disorientation) Refers to the outlier movement behaviors for elders with dementia in outdoor environments, which follow either the wandering pattern, the deviating pattern, or both, as defined above.

As such, trip t_1, t_2, and t_3 in Fig. 5.6 are therefore disorientation trips with t_1 following a deviating pattern t_2 and t_3 and I following a wandering pattern.

5.2.3.3   Problem Statement

The problem of disorientation detection is defined as follows: suppose we have an elder's historical movement trajectory set $T = \{t_1, t_1, \ldots, t_n\}$, where n is the total number of trajectories stored in the trajectory set. Given one elder's ongoing trip trajectory t, the objective is to inform in real time whether t is a disorientation trajectory that follows the deviating or wandering pattern as defined above.

5.2.4   iBDD: Isolation-Based Disorientation Detection

In this section, first we present an overview of our proposed disorientation detection method; we then provide some definitions which allow us to describe the transformation process from an original GPS trajectory into a symbolized sequence of traversed symbols.

5.2.4.1   Overview

As shown in Fig. 5.7, the proposed isolation-based disorientation detection (iBDD) method comprises two main steps. In the first step, we split the city map of interest into grid cells of equal size to form a cell matrix named CellMatrix. We use the obtained cell matrix to transform all GPS trajectories within a given set of historical trajectories, T, into sequences of traversed cells via a symbolization process. Finally, we instantiate the graph model TG with transformed trajectories in T to generate a new valid trajectory set TG. In the second step, we present our iBDD detection

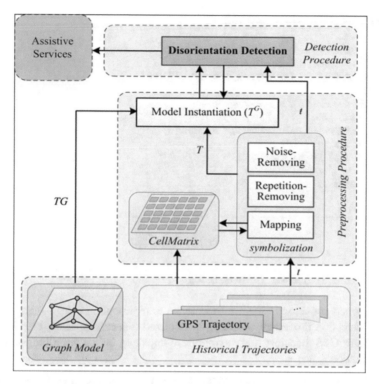

Fig. 5.7 Overview of the proposed iBDD method

method to examine whether an ongoing trajectory t is a disorientation trajectory. Specifically, we exploit the "few and different" properties of disorientation trajectories, and apply the isolation mechanism to detect disorientation trajectories.

Algorithm 5.1 summarizes the process of the iBDD method in detecting if an ongoing trajectory t should be classified as disorientation. The GPS points (e.g., p_i) of the test trajectory t are inputted one by one, and each of them is first transformed into a cell symbol by calling the symbolization procedure (which will be introduced in the next subsection), and then for a valid cell, the detection procedure (which will be introduced in Sect. 5.4) determines if the ongoing sub-trajectory t is a disorientation trajectory so far. If it is a disorientation trajectory, the algorithm outputs a value of $label = 1$.

Algorithm 5.1 The iBDD method

Inputs: $t = \langle p_1, p_2, \ldots, p_n \rangle$—a GPS trajectory
          CellMatrix—a city map matrix
          T^G—a set of symbolized trajectories
Outputs: *label* = 1 for disorientation, 0 for normal (default)
Process:
1: initialize T with T^G // T is a working set
2: **for** $i = 1$ to n **do**
3:        $(new, c) \leftarrow$ symbolization(p_i, *CellMatrix*)
4:        **if new** == 1 **then**
5:          $(l, os, T') \leftarrow$ detection(c, T)
6:          $T \leftarrow T'$
7:        **end if**
8: **end for**
9: *label* $\leftarrow 1$

5.2.4.2 GPS Trajectory Preprocessing

As mentioned above, we use GPS trajectories to represent an elder's outdoor trips, and the detection procedure works on symbolized trajectories. Here, therefore, we first define a GPS trajectory and a symbolized trajectory before describing a detailed transformation process.

Definition 2 (Elders' Mobility GPS Trajectory) A GPS trajectory, t, is a sequence of points $\langle p_1, p_2, \ldots, p_n \rangle$, where $p_i = (x_i, y_i)$ is the ith physical position in the form of (longitude, latitude). For a given $1 \leq i < j \leq n$, $t_{i \, : \, j} = \langle p_i, \ldots, p_j \rangle$ is called a sub-trajectory of t.

As GPS points are collected at a high sampling rate, many points often fall into a relatively small physical area, causing high computation cost and difficulty when comparing two trajectories. A symbolized trajectory is desirable to represent different instances of the same route consistently.

Definition 3 (Symbolized Mobility Trajectory) A symbolized trajectory, t, is a sequence of symbols $\langle c_1, c_2, \ldots, c_m \rangle$ obtained from t, where c_k $(1 \leq k \leq m \ll n)$ denotes the k-th physical area that t traversed.

The physical area c_k is generated by decomposing a city map of interest into a series of equal-sized cells according to a given size (e.g., 150 m $\times$ 150 m). Given a trajectory dataset of an elder, we are able to determine a valid range of the relevant city map according to the maximum (minimum) longitude and latitude value, and then decompose this area into a series of cells of the same size ($d \times d$). Finally, we name each cell and record it into a matrix of *CellMatrix* as follows:

Fig. 5.8 An illustration of
GPS trajectories and the
symbolized sequences

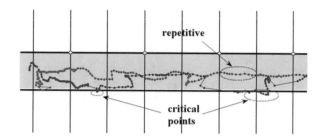

$$CellMatrix = \begin{bmatrix} c_{11} & \cdots & c_{1n} \\ \vdots & \ddots & \vdots \\ c_{m1} & \cdots & c_{mn} \end{bmatrix}. \tag{5.3}$$

Figure 5.8 shows two real GPS trajectories and the symbolized sequences (gray cells), where the GPS points falling into the same cells are replaced by the traversed cell from the *CellMatrix*. As depicted in Fig. 5.8, two trajectories traversing the same path may look very different if no trajectory preprocessing is done. The reasons for this are threefold. First, GPS trajectories usually have a high sampling rate (e.g., about 1–5 s or 5–10 m per point in our datasets). The high sampling rate results in more than one GPS point being recorded when a person traverses a cell. An elder walking the same path at different speeds will thus produce a different symbolized sequence of cells. Secondly, GPS signals are susceptible to surrounding conditions, such as bad weather and tall buildings, which can create noisy data such as a point that will register as originating from a neighboring cell. Finally, it is not unusual to have a path that traverses the border between two cells, as shown by the "critical points." This can cause two trajectories to be represented as traversing different sequences of cells.

We handle these issues by adopting the following trajectory symbolization process, which, in real time, transforms an original sequence of GPS points into a sequence of traversed cells, taking into account the above three factors. GPS trajectory symbolization: Symbolizing a trajectory means first mapping each GPS point into the matching cell by searching the *CellMatrix*, and then filtering repetitive and invalid cells. This will result in a symbolized sequence of traversed cells for the original GPS point sequence. The algorithm specifically looks for (and removes) invalid cells—those corresponding to noisy GPS data and critical points.

In order to transform the GPS trajectory in real time, we take a preexamination and post-evaluation combined approach, integrating distance and distribution density into the algorithm (see Algorithm 5.2). The algorithm starts by comparing the current GPS point to the previous point. If the current GPS point lies in the same cell as previous one, then a repetitive cell situation is identified and the cell is ignored (lines 4–6 in Algorithm 2). When the ongoing trajectory enters a new cell, the distance between the current point and the previous point is examined to estimate whether it is a "noisy" point. This is because GPS noise signals often lead to significant distance deviation. If a point is too far from the previous point, it can be considered noisy and discarded.

Algorithm 5.2 Symbolization (p', *CellMatrix'*)

Process:
1:   $t \leftarrow t \vee p'$ // $t = \langle p_1, p_2, \ldots, p_n \rangle$
2:   $s_i \leftarrow \text{Map}(p_i)$
3:   $s \leftarrow s \vee s_i$ // $s = \langle s_1, s_2, \ldots, s_n \rangle$
4:   **if** s_i is same with s_{i-1} **then**
5:       $n \leftarrow n + 1$ // counting same cell
6:       $nP \leftarrow nP \vee p_i$
7:   **else**
8:       $c \leftarrow c \vee s_j$ // c is used to record valid cells
9:       **if** dist(nP) $\leq d/3$ & $n \leq d/5r$ or dist(p_i, p_{1-1}) $\geq 2r$ **then**
10:          $new' \leftarrow 0$
11:          **if** $c(\text{end}) == c(\text{end}{-}2)$ **then**
12:              $c \leftarrow c(1 : \text{end}{-}1)$
13:          **end if**
14:          $nP \leftarrow \{ \}$
15:      **else**
16:          $new' \leftarrow 1$
17:      **end if**
18: **end if**
19: return (new', $c(\text{end}{-}1)$)

Second, after the ongoing trajectory leaves a cell, the traversed distance and the number of points within that cell are used to recognize "critical points" (i.e., those that occur because the path is along the border of two cells). Such critical points usually have a short traversed distance and a small number of points. Thus, as a result, the point falling into c_4' is recognized as a noisy point while the ones in c_6' are critical points (as shown in Fig. 5.9).

The suitable distance and proper density that a normal cell has are determined empirically according to the size of the cell and the sampling rate of GPS data. Suppose that the size of a cell is d and the sampling rate is r; a distance between two adjacent points of more than r strongly suggests that the latter is a noisy point. At the same time, a very short distance traversed by a small number of points may imply a critical category for these points. Again, in Fig. 5.9, the distance between the point in c_4' and its predecessor is significantly large, and the points traversed within c_6' are very small with only two points in it. We label cells resulting from noisy and critical points invalid and remove them from the symbolized trajectory.

Algorithm 5.2 summarizes the detailed symbolization process for an incoming GPS point p' (assume that it is the ith point p_i). Firstly, the algorithm concatenates the predecessors with p_i to form a new ongoing trajectory $t = \langle p_1, p_2, \ldots, p_n \rangle$, and maps p_i into the matching cell s_i by using the following mapping function:

$$\text{Map}(p) = \left\{ c_{kl} \begin{array}{l} \text{if } |x - x_{kl}| \leq d/2 \text{ and } |y - y_{kl}| \leq d/2 \\ \\ \hspace{3em} c_{kl} = (x_{kl}, y_{kl}) \end{array} \right\}. \quad (5.4)$$

If it lies in the same cell as its predecessor, the algorithm counts and records p_i into nP (lines 5 and 6). Otherwise, the mapped cell s_i is concatenated with

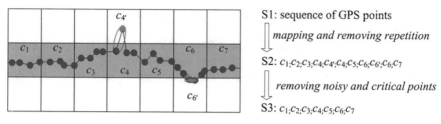

S1: sequence of GPS points

$\Downarrow$ *mapping and removing repetition*

S2: $c_1;c_2;c_3;c_{4'};c_4;c_4;c_5;c_6;c_{6'};c_6;c_7$

$\Downarrow$ *removing noisy and critical points*

S3: $c_1;c_2;c_3;c_4;c_5;c_6;c_7$

Fig. 5.9 An illustration of GPS trajectory symbolization process

c (recording non-repetitive cells), and the algorithm examines whether the preceding cell is an invalid cell (line 9). An invalid cell usually leads to cell repetition at its left and right neighbors as mentioned above, so it needs further processing (lines 11–13). For a normal cell, variable *new'* will be set as 1 (the default value is 0). Finally, the algorithm returns *new'* and $c(end-1)$ to the iBDD algorithm, with *new'* $= 1$ denoting a valid cell of $c(end-1)$ (line 4 in Algorithm 5.1). Specifically, dist(nP) means the traversed distance by all the points in nP, which records all points falling into a cell, and dist(p_i, p_{i-1}) is the distance between point p_i and p_{i-1}.

Note that the symbolization algorithm described in Algorithm 5.2 acts as a sub-procedure called by the iBDD algorithm for online detection of disorientation trajectories. For achieving a symbolized historical trajectory set, T, we only need to record the sequence of cells (symbols) in c for every GPS trajectory by simply adjusting the output of Algorithm 5.2.

The last stage in data preprocessing is to instantiate the defined graph model *TG* with the symbolized trajectories in T to obtain a new set TG. We first label all semantic cells that relate to the identified semantic places. Then we index all trajectories in T by counting not only the trajectories that contain exactly the home cell as their source and destination, but also the ones that traverse any other two semantic cells. This is because in a real situation, not every GPS trajectory (hence the transformed one) exactly starts at and ends in the home cell.

5.2.5 Disorientation Trajectory Detection Algorithm

This section introduces the disorientation detection process in detail. In order to test if an ongoing sub-trajectory is disorientation, we exploit its support degree in historical trajectories combined with the defined outlier patterns. Therefore, this section begins with some related definitions, and then presents the detection algorithm.

5.2.5.1 Preliminaries and Definitions

Definition 4 (Support, Support Set, and Support Degree) For two given trajectories $t_i \neq t_j \in T$, we say t_i supports t_j (or t_j is supported by t_i) if and only if t_j is a sub-trajectory of t_i. The set consisting of all symbolized trajectories in set T that support t_j is called support set of t_j:

$$T_{\text{supp}}^{t_j} = \left\{ t_i | t_i \in T \wedge t_i \text{ supports } t_j \right\}, \tag{5.5}$$

and the proportion of $T_{\text{supp}}^{t_j}$ in T is called support degree of t_j:

$$\text{supp}(t_j, T) = \left| T_{\text{supp}}^{t_j} \right| \times \frac{1}{|T|}. \tag{5.6}$$

For example, assume that $T = \{t_1, t_2, t_3\}$ is a set consisting of three trajectories $t_1 = \langle c_1, c_2, c_3, c_4 \rangle$, $t_2 = \langle c_1, c_2, c_3, c_4, c_5 \rangle$, and $t_3 = \langle c_2, c_3, c_4, c_5 \rangle$. Given two test trajectories $t' = \langle c_1, c_2, c_3 \rangle$ and $t'' = \langle c_2, c_3, c_4 \rangle$. According to Definition 4, t' has a support set of $\{t_1, t_2\}$ and support degree of 2/3 ($\approx$66.67%), because there are two trajectories t_1 and t_2 in T that support t' (i.e., t' is a sub-trajectory of both t_1 and t_2). Similarly, the support set and support degree for t'' are $\{t_1, t_2, t_3\}$ and 3/3 (100%), respectively.

As demonstrated by the above example, for a given trajectory set T, different test trajectories often obtain different support sets, and hence have different support degrees. In practice, it is very rare to obtain a support degree of 100% for any given trajectory. Because in real life, the set T consists of trips to different semantic places, and hence different traversed paths from one place to another. Consequently, we need to define a θ-support trajectory as follows.

Definition 5 (θ-Support) For a given constant $0 \leq \theta \leq 1$, a trajectory t_j is called θ-support with respect to T if its support degree $\text{supp}(t_j, T) \geq \theta$.

A θ-support trajectory is regarded as a normal trajectory with respect to the given constant threshold θ. Again, if θ is set as 0.5, t' and t'' are all normal trajectories in terms of T as shown in the above example. However, only t'' is normal if we assign 0.8 to θ. If a trajectory t_j has support degree $\text{supp}(t_j, T) \geq \theta$, then there are at least θ% trajectories in T that support t_j; otherwise, t_j is considered as "few and different."

It is obvious that threshold θ is crucial in determining whether a trajectory is normal or not. A higher value for θ implies a higher number of similar trajectories in a set. In real life, however, an elder's historical trajectory set T is actually divided into different clusters, with each cluster comprising trajectories to different semantic places. In most cases, we can even further partition a specific cluster into several different sub-clusters according to the traversed paths. Consequently, each (sub-)cluster has a different distribution density of trajectories, which will vary a lot for different semantic places. For instance, for a specific elder, the number of trajectories to grocery store may be greater than to daughter's home. Thus, θ should be determined depending on the elders' mobility history (i.e., recordings of historical trips) and should not be too large.

Definition 6 (Outlier Point and Outlier Score) An outlier point is a non-θ-support cell, where the trajectory either deviates from normal trajectories or changes direction repeatedly. The number of the found outlier points is chosen to be the outlier score (OS) for a disorientation trajectory.

According to Definition 1, a disorientation trajectory contains at least three continuous deviating or looping points. However, if a trajectory contains outlier point(s), no matter how big the number is, it is no longer θ-support according to Definitions 4 and 5. Since it is possible for an elder to recover and get back on track, an appropriate strategy is needed to break off the ongoing trajectory to ensure that the subsequent sub-trajectories can be considered θ-support. This is done depending on the types of outlier points: (1) deviating point, break into a sub-trajectory after the outlier point, and (2) looping point: break into a sub-trajectory before the previous point of the outlier point.

5.2.5.2 Disorientation Detection Algorithm

Algorithm 5.3 summarizes the detailed detection process for an ongoing trajectory $t = \langle p_1, p_2, \ldots, p_k \rangle$, where we suppose that p_k is the newly arrived GPS point, and it is mapped into a valid cell c_i. Additionally, we assume that t is considered a normal trajectory before the arrival of p_k; i.e., we have a symbolized ongoing sub-trajectory $t = \langle c_1, c_2, \ldots, c_{i-1} \rangle$, where $c_l (1 \leq l \leq i \ll k)$ are traversed cells by GPS trajectory t.

With the newly arrived cell c_i, the algorithm first concatenates $\langle c_1, c_2, \ldots, c_{i-1} \rangle$ with c_i to obtain a new sequence $t = \langle c_1, c_2, \ldots, c_{i-1}, c_i \rangle$, and then creates a support set for t by mining all trajectories in working set T_{ws}. After computing support degree, c_i will be recognized as an outlier point if t is not supported in terms of the given threshold constant θ (in line 4–12). If c_i is a looping outlier point, t will be broken at c_{i-1} (i.e., $t = \langle t(\text{end}-1), t(\text{end}) \rangle = \langle c_{i-1}, c_i \rangle$). If c_i is a deviating outlier point, t will be emptied (i.e., $t = \langle \rangle$). R_l and R_d are two variables used to count found outlier points of looping and deviating patterns, respectively. In both these cases, working set T_{ws} will be reset with the original set TG for a non-θ-support (sub-) trajectory (like t at the present moment).

Algorithm 5.3 Detection (c', T_{ws})

Process:

1: $t \leftarrow t \vee c'$ // $t = \langle c_1, c_2, \ldots, c_i \rangle$
2: search T_{ws} with t to obtain a support set T_{supp}^t
3: compute $\text{supp}(t, T_{ws})$ according to Eq. (5.6)
4: **if** $\text{supp}(t, T_{ws}) < \theta$ **then** // θ is a threshold constant
5:     **if** $t(\text{end}) = = t(\text{end} - 2)$ **then**
6:        $t \leftarrow \langle t(\text{end} - 1), t(\text{end}) \rangle$ // breaking at the place before $t(\text{end} - 1)$
7:        $R_l \leftarrow R_l + 1$ // counting looping outlier point
8:     **else**
9:        $t \leftarrow \langle \rangle$ // emptying t
10:       $R_d \leftarrow R_d + 1$ // counting deviating outlier point
11:    **end if**

(continued)

```
12:      reset T_ws with the original set T^G
13:   else
14:      T_ws ← T^t_supp
15:   end if
16:   if R_d or R_l ≥ 3 then
17:      l' ← 1 // with initial value of 0
18:      os' ← os' + 1 // with initial value of 0
19:   end if
20:   return (l', os', T_ws)
```

However, if c_i is detected as a normal point, there is only an update operation for T_{ws}: it will be reset with the support set of t (line 14). The use of the working set T_{ws} aims to improve the detection efficiency by limiting the search only within the θ-support trajectories in the previous step. So, an updated T_{ws} is returned by Algorithm 5.3 (line 20), corresponding to the one in line 5(6) in Algorithm 5.1.

If three outlier points have been found, t will be recognized as a trip with disorientation behavior (line 17 in Algorithm 5.3, and line 9 in Algorithm 5.1), and all remaining cells contribute outlier score 1 per cell according to Definition 6.

From the above detection process we can see that it is often insufficient to detect a disoriented trajectory by only mining support degree of an individual point; in other words, we need to examine the support degree of the ongoing trajectory $t = \langle c_1, c_2, \ldots, c_{i-1}, c_i \rangle$ instead of that of c_i. This is because a looping outlier point itself is still a θ-support cell occurring in the normal path that is traversed by the preceding cells; it is an outlier because the trip reverses direction from the prior cell (i.e., c_{i-1}). However, to obtain a support set for a long (sub-)trajectory by mining all trajectories in T_{ws} is obviously time consuming, causing lower performance for the detection system. We handle this issue by using a working trajectory t_{ws}, which only records the last three cells of a long (sub-)trajectory. The rationale for this is based on the following observation.

Observation 1 An ongoing trajectory $t = \langle c_1, c_2, \ldots, c_{i-1}, c_i \rangle$ is θ-support, if and only if sub-trajectory $\langle c_{i-2}, c_{i-1}, c_i \rangle$ is θ-support. On the contrary, $\langle c_{i-2}, c_{i-1}, c_i \rangle$ is the shortest non-θ-support sub-trajectory.

The necessity for Observation 1 is obvious, because any sub-trajectory of a θ-support trajectory is θ-support. So, a θ-support trajectory $t = \langle c_1, c_2, \ldots, c_{i-1}, c_i \rangle$ necessarily implies θ-support $\langle c_{i-2}, c_{i-1}, c_i \rangle$. For sufficiency, we first assume that $t = \langle c_1, c_2, \ldots, c_{i-1}, c_i \rangle$ is not θ-support. At this time, for any ordered sub-trajectory extracted from t, it is not θ-support if and only if it contains c_i. This is because $\langle c_1, c_2, \ldots, c_{i-1}, c_i \rangle$ is θ-support, or else it has been broken somewhere. Thus, the candidates are $\langle c_i \rangle$, $\langle c_{i-1}, c_i \rangle$, $\langle c_{i-2}, c_{i-1}, c_i \rangle$, and so on. As mentioned previously, however, both $\langle c_i \rangle$ and $\langle c_{i-1}, c_i \rangle$ may still be θ-support if t is detected as a looping trajectory. Therefore, $\langle c_{i-2}, c_{i-1}, c_i \rangle$ is the only shortest non-θ-support sub-trajectory.

Consequently, for a long (sub-)trajectory, we only use the last three elements to search trajectory set T_{ws} to obtain its support set, which is able to further improve detection efficiency. As such, the line 2 in Algorithm 5.3 is therefore modified as follows:

2: **if** $\text{len}(t) \geq 3$ **then**
  search T_{ws} with $\langle t(\text{end-2}), t(\text{end-1}), t(\text{end}) \rangle$ to obtain a support set
  else

Finally, we briefly explain why a trip that contains one or two outlier points is still categorized as a normal trajectory:

- For a looping pattern, an outlier point denotes a change of moving direction. One change may be caused by the fact that the elder changes his/her mind, while two changes may correspond to the case that an elder forgot something (first change) so he/she headed back, and he/she continued the trip (second change) after picking up the thing. Both are quite common in human daily life, so they are considered as normal trajectories.
- For a deviating pattern, an outlier point is one that is not on a normal trajectory. While one or two such points may simply be a small intended deviation, the larger the number of outlier points, the larger the deviation from normality. Since an assistive disorientation detection system should be robust to noise, small deviations should be tolerated. In our symbolized representation of cell sequences, a cell has a size of 150 m × 150 m, allowing a deviation of up to 300 m before being flagged as a possible disorientation event.

5.2.6 Performance Evaluation

In this section, we provide an empirical evaluation of the proposed approach using a real-world GPS dataset, which consists of more than 160 individuals' GPS traces released by Microsoft Research Asia [67, 68]. For more details about the experimental setup, please refer to the article [69].

5.2.6.1 Visualization

A visualization of the detection results is shown in Fig. 5.10, where both the manually labeled outlier trajectories and the outlier trajectories detected by iBDD are depicted. The outlier trajectories plotted in Fig. 5.10c are those detected by the iBDD method with the highest score. The number of these trajectories is equal to that of manually labeled trajectories in Fig. 5.10b, but the IDs are not necessarily identical. According to Fig. 5.10, we can see that the detected results are very similar to the manually labeled trajectories, indicating that iBDD is effective in detecting disorientation trips. Likewise, the iBDD method achieves similar results using other individuals' GPS datasets.

In Fig. 5.11, we present the detection results for two categories of outlier trajectories (both wandering and deviating patterns), where the red curves denote

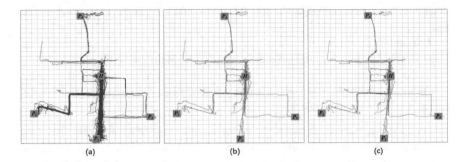

Fig. 5.10 An illustration of disorientation detection. H denotes center location and Pis are semantic places. (**a**) All trajectories, (**b**) manually labeled outlier trajectories, (**c**) iBDD-detected outlier trajectory

Fig. 5.11 An illustration of outlier trajectories detected by iBDD. op_1, op_2, and op_3 are the first outlier points in each disorientation trajectory

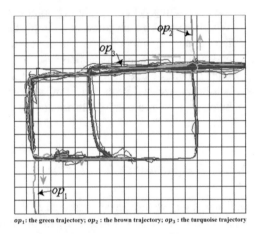

op_1: the green trajectory; op_2 : the brown trajectory; op_3 : the turquoise trajectory

the outlier parts of all three depicted trajectories, and op_1, op_2, and op_3 denote the starting points of the outlier trajectories. Specifically, a trajectory is not recognized as an outlier immediately after an outlier cell is found. Rather, a decision is made after two outlier cells (in a deviating pattern) or three changes of direction (in a wandering pattern) have been detected. The thick part of the wandering trajectory (depicted with the light blue line) is the loops, depicting a change of moving direction at least three times.

5.2.6.2   Quantitative Evaluation

We quantitatively evaluate the proposed iBDD method by calculating its AUC values for each chosen dataset. As shown in Table 5.3 (θ is set to 0.1), a value between 120 and 180 m for d obtains almost the same AUC values for all chosen datasets, with a detection rate of over 95% and a false-positive rate of less than 3%. However, if d is set to a value of less than 100 m or more than 200 m, the obtained

Table 5.3 The AUC values of iBDD with different values of cell size d when $\theta = 0.10$

d (m)	T-1	T-2	T-3	T-4	T-5	T-6	T-7	T-8	T-9	T-10
80	0.8872	0.8753	0.8990	0.8999	0.8824	0.9001	0.9104	0.9200	0.9006	0.8942
120	0.9903	0.9911	0.9921	0.9933	0.9875	0.9941	0.9940	0.9954	0.9960	0.9959
150	**0.9972**	**0.9995**	**0.9962**	**0.9968**	**0.9967**	**0.9997**	**0.9977**	**0.9974**	**0.9994**	**0.9992**
180	0.9912	0.9931	0.9929	0.9927	0.9915	0.9939	0.9953	0.9955	0.9959	0.9969
220	0.9012	0.8994	0.8999	0.9100	0.8971	0.9033	0.9320	0.9297	0.9119	0.9022

The best performance is obtained while the cell size d is set as 150 m

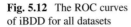

Fig. 5.12 The ROC curves of iBDD for all datasets

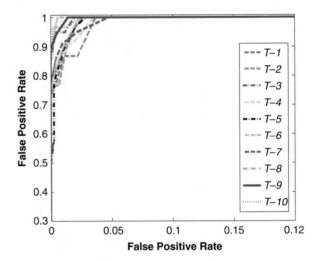

AUC value will decline significantly. The reason for this phenomenon might be that continuous GPS points of a normal trajectory (e.g., a walking trajectory) will be mapped into different cells if the value of d is relatively small, leading to a high false-positive rate. At the same time, loop-like movements might be mapped into the same cell in case a large d is used, resulting in a low detection rate.

As the target users of the proposed iBDD method are elders with impaired cognition, a disorientation trajectory must be detected in a timely manner before the elder deviates too far away from one of the normal paths. Thus, a value of 150 m for d is not only algorithmically feasible, but also acceptable for real applications. In this case, an elder will travel about 300 m (i.e., two cells) before his/her disorientation behavior can be detected. Based on this computed value for d, we depict the ROC curves of iBDD on all datasets in Fig. 5.12. It can be seen that the proposed method is able to achieve a high detection rate while keeping a low false-positive rate.

Another important parameter of the iBDD method is the threshold constant θ, which should be determined according to the distribution of trajectories. Generally, an even distribution of trajectories in a given dataset corresponds to a large θ. In a real-world application, trajectory distribution is directly related to the type and number of individual's semantic places. For example, for many elders, there will be significantly more trajectories to community centers than to shopping malls. We performed experiments using different θ (including 0.05, 0.10, and 0.20), and the result is shown in Table 5.4, which suggests that $\theta = 0.10$ is suitable for evaluating the proposed method with the chosen datasets. Specifically, dataset $T - 6, T - 7$, and $T - 8$ obtained much higher AUC values than other datasets when θ is set to 0.20; the reason is that there are only two semantic places in these three datasets and trajectories in these datasets are relatively neat.

Table 5.4 The AUC values of iBDD with different values of θ when $d = 150$ m

θ	T-1	T-2	T-3	T-4	T-5	T-6	T-7	T-8	T-9	T-10
0.05	0.9979	0.9991	0.9967	0.9971	0.9971	0.9998	0.9981	0.9978	0.9995	0.9997
0.10	**0.9972**	**0.9995**	**0.9962**	**0.9968**	**0.9967**	**0.9997**	**0.9977**	**0.9974**	**0.9994**	**0.9992**
0.20	0.7731	0.7912	0.8005	0.7889	0.7544	0.9210	0.9074	0.9309	0.8127	0.7999

With the chosen datasets, the best performance is obtained while the value of θ is set as 0.10

5.3   Human Computer Operation Recognition Based on Smartphone[3]

5.3.1   Introduction

Human-computer operations such as document processing, spreadsheets, programming [70], and digital games [71] have become increasingly involved in our everyday life. These activities reflect interactions between the user and the computer, which can be exploited to facilitate context-aware services [72]. In particular, human-computer operations may reflect one's real-time status as well as specific desire or requirements. For example, if a user is working on a document or coding, it is very likely that he/she does not want to be disturbed, and the mode of his/her phone should be set to the silent mode. In contrast, when a user is chatting or playing games, he/she may want to be in a relaxed atmosphere, and light music may be recommended to him/her. Based on the user's operations, we can adapt the environment to different contexts, such as adjusting the volume of the phone ringing, or the light intensity in the room. Further, we can characterize the user with more information by using his/her operations on computer, e.g., a computer game enthusiast or a hardworking person, which is also helpful to provide personalized services. In addition, humans are usually absorbed in one computer operation [73] and do not notice the time. To keep a healthy lifestyle, sensing human-computer operation can help to set a personal reminder for taking a break or a new task. Moreover, recognizing these operations can be used to generate a personalized activity schedule and properly manage time for the user. Therefore, aiming at enabling all these context-aware services in a pervasive and nonintrusive manner, we propose to recognize human-computer operations based on keystroke sensing with smartphones.

In general, human-computer operations can be recognized by installing a keylogger in the computer. However, providing personalized services via a computer (e.g., adjusting the light intensity) is not very convenient compared with using a smartphone or other wearable devices. Specifically, prior work has extensively studied human activity recognition [74, 75] and context-aware sensing [76–79] by utilizing smartphones, ambient sensors, and wearable devices. For example, by using ambient sensors or wearable devices, prior research that mainly focused on characterizing a user's physical activities and body movements was able to recognize the refrigerator opening [80], eating [81], etc. With multiple sensors embedded, a smartphone can also be used to recognize a user's activities, e.g., brushing teeth [82], running [83], and computer keystrokes [84–87]. In this chapter, we aim to recognize human-computer operations by using a smartphone. In particular, we do not adopt the technologies of ambient sensors or wearable devices mainly due to the

³Part of this section is based on a previous work: Z. Yu, H. Du, D. Xiao, Z. Wang, Q. Han and B. Guo, "Recognition of Human Computer Operations Based on Keystroke Sensing by Smartphone Microphone," in IEEE Internet of Things Journal, vol. 5, no. 2, pp. 1156–1168, April 2018. DOI: https://doi.org/10.1109/JIOT.2018.2797896

following factors. The deployment of ambient sensors can be labor intensive and costly, and wearable devices are not always convenient or available. In addition, there exists some work related to recognition of human-computer operations by using smartphones. For instance, keystrokes can be identified using the audio features of each keystroke on the keyboard [84, 85]. However, we cannot determine the type of operation only based on individual keys. Building on the existing studies, we further explore the feasibility of using smartphones for human-computer operation recognition.

In this chapter, we capture a user's keyboard inputs and recognize human-computer operations by utilizing a single microphone in smartphones. We face several interesting challenges. First, there are various kinds of human-computer operations (e.g., writing documents or coding) that are rich in semantics, but the data collected using smartphone sensors is quite limited and may not show much difference between different operations. Therefore, obtaining the semantic information of human-computer operation based on sensor data is essential yet challenging. Second, distinct features that can be used for recognition of human-computer operations are still to be explored.

To tackle these challenges, we sense the content of a user's keyboard input, and recognize four human-computer operations, i.e., coding, writing documents, chatting, and playing games, which are commonly seen at work and entertainment. The effective recognition inspires our future exploration taking more human-computer operations into consideration. In summary, the contributions of this chapter are threefold.

- We capture the audio signals from keyboard input by utilizing the single microphone embedded in most commodity smartphones, and recognize the input words. Specifically, the keystrokes are first identified by the extracted acoustic features in frequency domain. To further correct the identified words, we then apply the methods in natural language processing (NLP) to obtain N-Gram-based candidate word set, and select the right word from it by proposing an adjacent similarity matrix algorithm.
- We extract the features on semantics and audio signal, and further employ the AdaBoost algorithm for user's computer operation recognition, i.e., chatting, coding, writing documents, and playing games. In particular, based on the corrected words, we propose lexicon preference and semantic rationality to characterize each human-computer operation in semantics. Meanwhile, based on the input audio signals, input rate and audio energy are also used for human-computer operation recognition.
- We conduct experiments in realistic environments to validate the effectiveness of the proposed method. Specifically, we consider two types of keyboards and input modes. Systematic analysis has been conducted and presented on the performance of the proposed features and algorithms.

5.3.2 Related Work

5.3.2.1 Human-Computer Operation Recognition

Human-computer operations record what a user has done on the computer, which has been explored by existing studies and software. For instance, data from computer games and psychomotor measurements associated with keyboard entries and mouse movement can be monitored to infer a user's cognitive performance [88]. Meanwhile, discovering usage pattern from Web data is also commonly used to understand and better serve the needs of Web-based applications [89]. These studies have mainly focused on the understanding of one type of human-computer operations, whereas there is not a comprehensive study for user's various activities on the computer. In addition, ActivTrak (https://activtrak.com/) monitors people's online activities to ensure greater productivity and safety. It works on a computer and collects application title bars, page titles, URLs, and screenshots. This type of application cannot provide more context-aware services in a pervasive and nonintrusive manner for the users, e.g., adjusting the mode of the phone ringing or the light intensity of the room. As many powerful sensors are now embedded in the smartphone, we want to monitor a user's various computer operations by utilizing the smartphone, which can facilitate the development of personalized services, e.g., personal reminder, phone mode setting, and indoor light adjustment.

5.3.2.2 Human Activity Identification Using Smartphones

The fast development of sensor-enriched smartphones provides an opportunity that human activity can be sensed by smartphones, ranging from personal movements to crowd behaviors [90–93]. For example, accelerometer embedded in smartphones is capable of characterizing human's movements, e.g., standing, walking, and running [94–96]. Gyroscope sensor is the most sensitive sensor for measuring the heart rate [97]. Meanwhile, the microphone has been widely used to sense a user's activity. For instance, by collecting the audio from the phone's microphone, it is possible to recognize a user's activity, e.g., listening to music, speaking, and sleeping [98, 99]. By using accelerometer and microphone in a smartphone, a system can monitor running rhythm to help users better understand their running process [83]. By using the built-in microphone in the smartphone, SymDetector [100] detects the sound-related respiratory symptoms of a user, such as sneeze or cough. These studies offer related techniques and algorithms on smartphone sensing, which support and inspire this chapter.

5.3.2.3   Keystroke Recognition

Keyboard input recognition is the basis of our proposed human-computer operation recognition. There are some studies on keystroke identification by using sound-wave range measurement [101, 102] and wireless signals [103, 104]. In particular, the method of sound-wave range measurement requires two or three smartphones, and the technique of wireless signal sensing has stringent requirements for the environment. We hence do not use any of the two methods in this chapter.

In addition, the acoustic signal generated by each keystroke is unique, and hence can be regarded as a fingerprint feature [85]. By utilizing these observations, keystrokes were recognized with a supervised algorithm [105]. Trading off training needs for accuracy through a dictionary-based approach was investigated [106]. Moreover, combined with the acoustic model of multipath fading, the keystroke hit in a solid surface was also recognized [86]. Using the insights gained from these studies, the keyboard input can be effectively recognized by utilizing a microphone built in a smartphone based on the fingerprint features of the keystroke audio. Different from existing studies, we extract 400 features in frequency domain for the keystroke identification. Also, prior research only studied the keystroke recognition, whereas we aim to recognize human-computer operations, which cannot be achieved only by using separate keystrokes identified. Therefore, this chapter advances existing work on keystroke sensing.

5.3.3   System Overview

With the single microphone in a commodity smartphone, we record the acoustic signal from keystrokes and recognize the user's computer operations. The proposed framework consists of three major components: (1) keystroke identification, (2) word correction, and (3) human-computer operation recognition, as shown in Fig. 5.13.

We first preprocess the acoustic signals from keystrokes by denoising and key separation. We then identify the keystrokes by extracting features in frequency domain and correct the input words. Specifically, based on the N-Gram model in NLP, we propose non-repeating word segmentation-based N-Gram distance to select the candidate word set. Utilizing the characteristics of audio signals, we design an algorithm based on adjacent similarity matrix to recognize the right word. In particular, the position relations of the letters in a word are redefined to obtain the constraint set, and then the constraint subset method is applied to eliminate the effects of error conditions. After that, we extract the characteristics on semantics and audio signals, and further acquire the activity segment. The AdaBoost algorithm is then applied to recognize the user's computer operations.

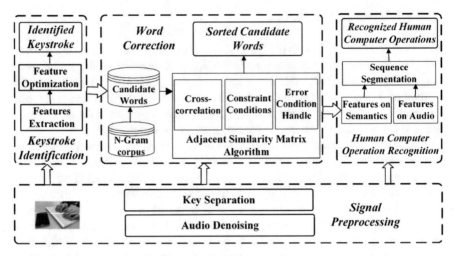

Fig. 5.13 Overview of the proposed approach

5.3.4   Keystroke Identification

When typing different keys on the keyboard, distinct audio signals are generated. For example, the sound of typing the space bar can be easily distinguished from other keystrokes. Generally, these differences are caused by various reasons, e.g., the unique structures of some keys, the surrounding context of the keys in the keyboard (i.e., the structures of ambient keys), and the distance between the key and human ears [85, 101]. Therefore, it is possible to differentiate different keystrokes by leveraging acoustic features. A keystroke consists of two stages: (1) press and (2) release. If the release of a keystroke overlaps with the press of the next keystroke, it is difficult to separate them. Therefore, to better address keystroke identification, we make two assumptions. First, there is no input from key combinations, e.g., "ctrl + O." Second, a key is pressed only when the previous one has been released.

In this chapter, we identify keystrokes by utilizing the single microphone embedded in smartphones. The audio waveform of the keystroke is short with bursts, as shown in Fig. 5.14. When a key is tapped continuously, there are frequent fluctuations in the waveform, resulting in multiple peaks in zero-crossing rate (ZCR).

However, the effects of noises are obvious that also generate peaks in ZCR. Also, the sound vibration of a keystroke in push and release stages is evident. As shown in Fig. 5.15, there is a silence period between the push and release. Similarly, there are also two stages when we press a key and do not release it, i.e., touch and hit. When we touch the surface of the key and then press down, it shows two peaks with a small silence period in the audio waveform of the push. We will exploit the characteristics of the audio waveforms in different stages to recognize keystrokes.

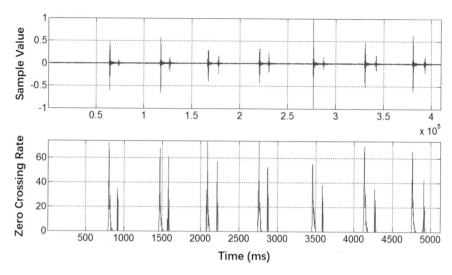

Fig. 5.14 Audio waveform and ZCR when "V" is pressed several times

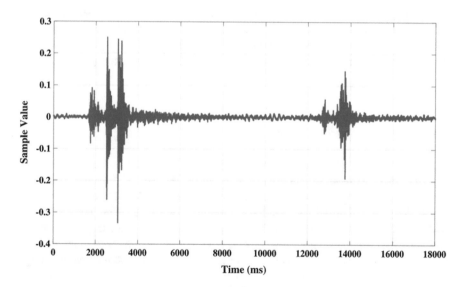

Fig. 5.15 Audio waveform of the keystroke

5.3.4.1   Data Preprocessing

To eliminate the effects of noises on ZCR and obtain clear features of the keystrokes, we first filter the original audio signals by utilizing the Wiener filtering algorithm [107]. We then adopt endpoint detection to split each keystroke event, a widely used

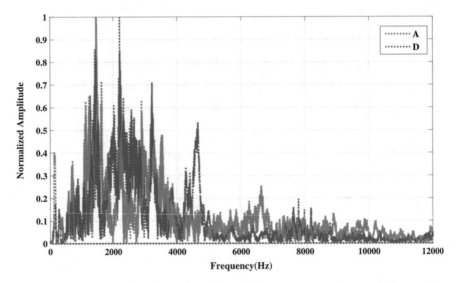

Fig. 5.16 Spectrum of the keystroke "A" and "D"

approach in speech detection [108]. In particular, the audio of a keystroke usually has a short duration with high vibration frequency and large amplitude, leading to clear peaks in short-term energy and short-time ZCR. Therefore, we use short-term energy and short-time ZCR as thresholds to achieve double-threshold endpoint detection.

5.3.4.2   Keystroke Identification

Audio waveform spectrums corresponding to different keystrokes are distinct. Figure 5.16 presents the spectrums of keystrokes "A" and "D," with a frequency range of 0–12 kHz and the normalized energy. We observe that the frequency bands where the peaks of energy are located are distinct for "A" and "D" between 0 and 8 kHz. Moreover, when the frequency is higher than 8 kHz, the energy is very low and the difference between energy peaks is not obvious. In other words, there is less information in the high-frequency area. Therefore, we utilize the features of energy peaks in the range of 0–8 kHz to recognize keystrokes.

We first obtain the spectrum of each keystroke audio by applying fast Fourier transform (FFT), where the frequency range is set to 0–8 kHz. In particular, a 2 ms Hamming window is applied in FFT. Afterwards, a frequency window of 20 Hz is used to analyze the energy in different frequency bands. Specifically, the spectrum is divided into 400 segments, and we compute the average energy as the feature in each segment. In addition, the features are extracted in different stages of a keystroke, respectively, i.e., the push and the release. Using the 400 features in frequency domain, we apply a decision tree algorithm to recognize which key is pressed for a specific keystroke audio.

5.3.5 *Word Correction*

There may exist errors in keystroke identification; for example, a letter in a word is not correctly identified, which may lead to accumulated errors for determining the input series of human-computer operations. Therefore, we further refine the identified input series via word correction. Generally, words can be segmented by the space bar or enter key, which can be accurately identified with the method described above. We then map each separated letter sequence to the most appropriate word in a candidate word set to improve the results of keystroke identification.

5.3.5.1 N-Gram-Based Candidate Word Set Determination

With the identified keystrokes, the candidate word set is determined based on fuzzy matching, depending on the distance between the identified letter sequence and the word in corpus. Hence, the metric for word distance has great impact on the matching accuracy. Popular metrics include cosine similarity, edit distance, etc. In this chapter, we utilize the N-Gram distance in N-Gram model and propose $NRNG_{Dist}$ distance as the metric.

N-Gram model [109] is widely used to predict or evaluate the validity of a sentence based on some corpus in NLP. It can also evaluate the difference of two strings, which is a popular method in fuzzy matching. For a word s of length M, N-Gram model splits it into a list of substrings, each of which has N letters and is called a Gram. With an offset of one letter, $M - N - 1$ Grams can be generated. The N-Gram distance between two words is defined as the number of same Grams. This method can effectively measure the distance of two words that have the same lengths. However, for words with different lengths, it is not satisfactory. For example, the N-Gram distance of "boy" and "boyfriend" is equal to the number of Grams that boy has, but we cannot conclude that the two words are the same.

To measure the distance of words with different lengths, we propose non-repeating word segmentation-based N-Gram distance, named $NRNG_{Dist}$, as follows:

$$NRNG_{Dist} = |\, G_N(s)\,| + |\, G_N(t)\,| - 2 \times |\, G_N(s) \cap G_N(t)\,|, \qquad (5.7)$$

where $GN(s)$ is the N-Gram set of string s, $|GN(s)|$ is the size of $GN(s)$, and N is normally set as 2 or 3.

$NRNG_{Dist}$ not only considers the number of same Grams, but also takes into consideration the difference of lengths. A higher similarity of two words corresponds to a smaller $NRNG_{Dist}$. If the value of $NRNG_{Dist}$ is zero, the two words are the same. For the input word or letter sequence iw, if the $NRNG_{Dist}$ between iw and the word cw in corpus is smaller than a distance threshold ε, then cw is selected as a candidate word. Note that we put the original input letter sequence in the candidate set, as iw can be a new word that is not in the corpus.

5.3.5.2   Word Recognition by Using the Adjacent Similarity Matrix Algorithm

As the N-Gram-based candidate set can be quite large, we further introduce several constraints to reduce the set. We observe that the audio of a keystroke is closely related to the key's physical position [85] and the audio waves are similar when pressing the same or physically close keys. Therefore, by utilizing the input audio, we can obtain the position similarity of letters in a word, and use such position similarities as the constraint.

We first use cross-correlation to describe the relations between any two letters in a word w, which is widely adopted in acoustic distance measurement [101]. Based on the time series of audio from the input word that contains N letters, a correlation matrix $X_{\text{corr}} = (\text{Sim}_{ij})$ is obtained, where Sim_{ij} represents the correlation between the ith letter and jth letter of w. However, the correlation is hard to reach 1, even if the same key is hit twice. Therefore, it is difficult to determine the relation of keys from cross-correlation. Moreover, the correlation matrix consists of $N \times (N - 1)/2$ constraints, the computation complexity of which grows exponentially with N. What exacerbates the problem is that an error condition may lead to incorrect recognition. To address these problems, we propose an adjacent similarity matrix algorithm, as shown in Algorithm 5.4.

Algorithm 5.4 Adjacent similarity matrix algorithm

Input: Candidate word set $\Psi = l_1, l_2, \ldots, l_z$ and audio signal of word $l = k_1 k_2 \ldots k_n$
Output: The word l_{goal} satisfying maximum constraint subsets
Process:
1:   **for** each $k_i \in l$ **do**
2:        $Rank(i, j) \leftarrow$ Xcorr;
3:   **end for**
4:   $\Pi \leftarrow Rank(i, j)$;
5:   set p and m;
6:   $\Gamma \leftarrow \Lambda, \Pi$;
7:   $W \leftarrow \Psi, \Pi$;
8:   $\Omega = (\omega_1, \omega_2, \ldots, \omega_k) = \text{Transform}(\Gamma, W)$;
9:   $maxm = 0$;
10: **for** ω_i in Ω **do**
11:       **if** $\|\omega_i\|_0 > maxm$ **then**
12:            $maxm = \|\omega_i\|_0$;
13:            $goal = i$;
14:       **end if**
15: **end for**
16: return l_{goal}

Specifically, based on the correlation matrix X_{corr}, we redefine the relation of two keys as follows. The diagonal values Sim_{ii} representing autocorrelation are set to 0 to avoid interference. For key K_i, we first sort the similarity between K_i and other keys in descending order. $Rank(i, j)$ denotes the ranking of $\text{Sim}(i, j)$ for K_i. If $Rank(i, j)$ is higher for K_i and $Rank(j, i)$ is higher for Kj, i.e., K_i and K_j are very similar, then there is a high possibility that K_i is the same as K_j. Utilizing these rankings, we redefine the relations between two keys as 5.8:

$$\begin{cases} \text{Rank}(i,j) + \text{Rank}(j,i) \leq 3 \Rightarrow K_i = K_j \\ \text{Rank}(i,j) + \text{Rank}(j,i) = 4 \Rightarrow K_i \simeq K_j \\ \text{Rank}(i,j) + \text{Rank}(j,i) = 5 \Rightarrow K_i \approx K_j \\ \text{Rank}(i,j) + \text{Rank}(j,i) \geq 6 \Rightarrow K_i \sim K_j \end{cases} \quad (5.8)$$

In fact, the word that satisfies all the constraints is usually not unique. We next further refine the constraints. There is a trade-off between the size of the constraint set and the accuracy of word determination. On the one hand, a comprehensive constraint set is too strict, since the right word cannot be determined with even just one false condition. On the other hand, a small constraint set can be satisfied by several words and a single condition cannot determine the correct word. Thus, we propose to use multiple subsets of the whole constraint set, and evaluate each subset to select words. Specifically, a candidate word set has been determined based on the $\text{NRNG}_{\text{Dist}}$ distance, and we want to select the correct one based on some constraints. By utilizing the relation between any two keys in a word, we can get a constraint set Π, where the relation of two keys is denoted as $\odot$. The subset of constraints is represented as $\odot$, and it denotes $\Lambda \subseteq \{K_i \odot K_j | i \leq N, j \leq N\} = \eta_1, \eta_2, \ldots, \eta_m$. The constraint matrix, denoted as $\Gamma = (\gamma nm)$, represents the conditions in each constraint subset. $\gamma nm = 1$ represents that the condition ηm is in the subset n, i.e., $\eta_m \in n$. For each condition $\eta = K_i \odot K_j$ in Λ or in Π, we traverse the candidate word set Ψ, and select words that satisfy η. The selected word set is denoted as Ψ^η. The constraint-word matrix $W = (w_{mk})$ represents the relation between the candidate word set Ψ and the constraint set Π, where $w_{mk} = 1$ means that the kth word l_k in Ψ satisfies the mth constraint in Π, i.e., $l_k \in \Psi^{\eta m}$. The words satisfying the constraint subset Λ can be represented as $\Psi^\Lambda = \cap \Psi^\eta$. The correct word l_{goal} is determined by sampled constraint subsets.

For a word consisting of N letters, the number of constraints and constraint subsets are $N \times (N - 1)/2$ and $2^{N \times (N-1)/2}$, respectively. We refine the word constraint via sampling all constraint subsets m times. Following the equal probability principle, a constraint subset will be selected by one sampling and will not be put back. After m sampling, m subsets are obtained, each of which has $N \times (N - 1)/2 \times p$ conditions. The values of p and m are empirically determined. Using the obtained m condition subsets, we analyze the times that the same word is selected, i.e., the number of subsets that the word satisfies. The more frequently selected word is more likely to be the right one. By combining the constraint matrix Γ with the constraint-word matrix W, we can obtain the relations between m constraint subsets and the candidate words, which is defined as a matrix $\Omega = (\omega_{nk})$ by using $\Omega = Transform$ $(\Gamma \times W)$. Specifically, the function $Transform()$ traverses and resets the matrix elements v_{ij} of $\Gamma \times W$. If the value of v_{ij} equals the size of i, then v_{ij} is set to 1, i.e., $\omega_{ij} = 1$; otherwise, it is set to 0, i.e., $\omega_{ij} = 0$. In this case, $\omega_{nk} = 1$ denotes

$l_k \in \Psi^{\wedge n}$. With this matrix, we can derive the selected times of the same word, which corresponds to the number of the value 1 in each column of Ω. Finally, the one with the most selected times is regarded as the correct word.

5.3.6   Human-Computer Operation Recognition

Based on the corrected word series, we characterize and recognize a user's computer operations. During a period of time, the audio of a sequence of keystrokes can be from various operations, e.g., chatting, writing documents, and playing games. The inputs of these human-computer operations have distinct features in semantics. For instance, "U" represents "you" in chatting, while this abbreviation is uncommon in documents [110]. The audio wave is also distinguishable. For example, the intensity of pressing keys is stronger when playing games [111, 112]. Therefore, we utilize these characteristics in audio wave and semantics to recognize four common human-computer operations, i.e., chatting, coding, writing documents, and playing games.

5.3.6.1   Semantic Features

Lexicon Preference: Different human-computer operations have distinct lexicons; for example, the keywords are more common in coding, and the operation keys such as "Q" and "W" are widely used when playing games. Based on the differences of the input words, we extract the lexicon preference to characterize different human-computer operations.

In this chapter, we first define the identification set for each human-computer operation, respectively, based on the Corpus of Contemporary American English (http://www.wordfrequency.info/). Specifically, the identification set of chatting is formed by the abbreviations commonly used in chatting (represented as φ_1); the identification set of writing documents is composed of the most frequently used 50 words (represented as φ_2); the identification set of playing games (represented as φ_3) contains the operation keys commonly used in games, such as "Q," "W," "E," and "R"; and the identification set of coding consists of the keywords and reserved words (represented as φ_4). Afterwards, the *Dice* coefficient is used to measure the distance between the input series S and the specific identification set of each human-computer operation. Thus, the preferences of S for each operation can be obtained, i.e., D_1, D_2, D_3, and D_4.

Semantic Rationality: In general, the inputs of documents are more rational than those of coding in semantics. Similarly, the inputs when playing games seem random, while the words for chatting are more casual than documents [35]. Therefore, we propose semantic rationality to describe human-computer operations.

In particular, we apply N-Gram model to evaluate the semantic rationality. In NLP, a sentence S is composed of any N words in any orders, and the occurrence probability of the sentence, $P(S)$, is then regarded as the semantic rationality. For example, if S1 = "he is eating" and S2 = "is eating he," then S1 has more clear semantic meaning, i.e., $P(S1) > P(S2)$. For an input segment $S = \{w_1, w_2, \ldots, w_m\}$ composed of m words, the rationality $P(S)$ can be obtained based on the chain rule, i.e., $P(S) = P(w_1, w_2, \ldots, w_m) = P(w_1)P(w_2|w_1)P(w_3|w_1, w_2) \cdots P(w_m|w_1, \ldots, w_{m-1})$.

However, we can observe that the value of $P(S)$ is 0 for most of the sentences, as the requirements are too strict. The computation complexity is also high. To address these issues, we use $N - 1$ order Markov hypothesis to optimize the probability, where the probability of w_i only depends on the previous $N - 1$ words, as follows:

$$P(w_i|w_1, w_2, \ldots, w_{i-1}) = P(w_i|w_{i-N+1}, w_{i-N+2}, \ldots, w_{i-1}). \qquad (5.9)$$

In NLP, N is usually set as 1, 2, or 3, and a $N - 1$ order Markov hypothesis corresponds to a N-Gram model. In this chapter, we adopt the 2-Gram model or bigram model, i.e., $N = 2$. Based on the conditional probability formula, the bigram model can be defined as

$$P(w_1, w_2, \ldots, w_m) = \prod_{i=1}^{m} P(w_i|w_{i-1}) = \prod_{i=1}^{m} \frac{P(w_i w_{i-1})}{P(w_{i-1})}. \qquad (5.10)$$

We observe that the determination of $P(w_{i-1})$ and $P(w_i w_{i-1})$ is of great importance, which relies on the huge amount of data in corpus. Fortunately, Google Books (https://books.google.com/ngrams), with a 0.2–0.3-s response time, provides free online inquiry of N-Gram, which is employed in this work.

5.3.6.2   Acoustic Features

Acoustic features are significant for the recognition of human-computer operation, which can be extracted from two aspects. First, the input rate is effective for recognition; for example, compared with programming, chatting is more casual and often done spontaneously, and hence more likely to have a higher input rate. Second, the energy of the audio, which can represent the power of pressing keys, is another piece of useful information. For example, when the users are playing games, they tend to press keys with more power [111, 112]. Therefore, we adopt the input rate v and the audio energy E to characterize the audio signals.

5.3.6.3   Human-Computer Operation Recognition

An input sequence may consist of multiple human-computer operations; that is, the entire time period can be divided into multiple nonoverlapping time segments and each segment is for one operation. Therefore, to accurately recognize user's computer operations, we need to segment the input series and recognize each suboperation. We recognize human-computer operations based on the operation segmentation. We first use a sliding window for time division. Specifically, to avoid breaking an input sentence, we regard each word as a point. In such a way, the audio segment st in each time window is composed of multiple words. Based on the four features defined, we can use a feature vector ft. = (DV, P, v, E) to represent st. If two segments belong to the same human-computer operation in consecutive time slots, then their features are more similar than those of other segments. Therefore, human-computer operations can be differentiated based on the similarity of adjacent feature vectors.

To obtain each operation segment, we propose a sliding window slicing algorithm based on the feature vector difference. In case that the time window is 3, the input series $X = w_1, w_2, \ldots, w_n$ can be represented as a set of triples $s_i = w_i w_{i+1} w_{i+2}$, and each triple corresponds to a feature vector $ft_i = (DV, P, v, E)$. Therefore, by setting the sliding step as 1, X can be denoted using the feature vector series $ft_1 ft_2 \ldots ft_{n-2}$. In other words, the segmentation of input series can be achieved by slicing the feature vector series. As we have illustrated, if the feature vectors in consecutive time belong to the same human-computer operation, the distance between them would be small. The distance of adjacent feature vectors can be measured by utilizing cosine similarity and Tanimoto coefficient.

Based on the distance between feature vectors, we can obtain the operation segment of the input series. An operation segment with a length of n will have $n - 2$ ft., and the average of ft_i is defined as the feature of this operation segment to be used for human-computer operation classification. In this chapter, we select three classification models (i.e., AdaBoost, SVM, and ZeroR) to infer the type of each segment of human-computer operation. AdaBoost is an adaptive boosting machine learning algorithm, which converges to a stronger learner by learning from the results of weak learners. SVM models the examples as points in space, and the examples of the separate categories are divided by a clear gap that is as wide as possible. ZeroR regards the majority classes as the classification results based on statistical rules of historical data. We will compare the performance of these three inference models in the experiments.

5.3.7   Performance Evaluation

We used two types of keyboards in the experiments, i.e., the 87-Key mechanical keyboard Dareu, and the membrane keyboard Rapoo E1050. Moreover, an iPhone 5s was utilized to record the keyboard inputs based on its single microphone. The sampling rate of microphone is 44.1 kHz. During the experiments, the phone and the keyboard are placed side by side, as shown in Fig. 5.17.

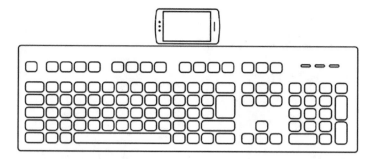

Fig. 5.17   Placements of the keyboard and the smartphone

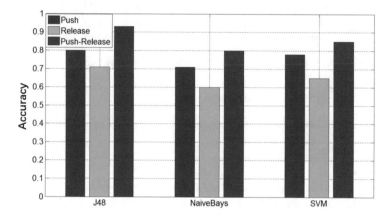

Fig. 5.18   Keystroke identification results by utilizing the features from different stages

5.3.7.1   Experimental Results of Keystroke Identification

To identify keystrokes, we extracted 400 features in the frequency domain, which are distributed in 0–8 kHz. We recorded the audio waveform of 20 keystrokes from each letter as well as some commonly used symbol keys. Three different classifiers (i.e., J48, Naive Bayes, and SVM) are trained based on features from only the push, the release, or both the push and release stages. Experimental results (as shown in Fig. 5.18) indicate that for each classifier, features extracted from the push stage play a more important role than features extracted from the release stage. The reason might be that the intensity of pushing keys is stronger releasing keys. Specifically, the best performance is obtained by using features from both stages. Accordingly, we can conclude that different keystrokes have distinct frequency bands where the peaks appear.

Furthermore, we used J48 to evaluate the keystroke identification performance when using different types of keyboards. The result is shown in Fig. 5.19; we find that the identification accuracy of the mechanical keyboard is much higher than the

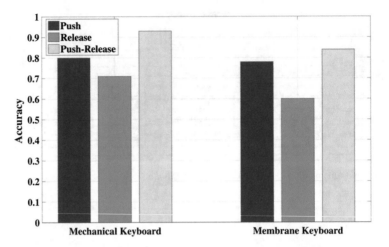

Fig. 5.19 Keystroke identification results by using two types of keyboards

Table 5.5 Results of human-computer operation partition

Similarity metric	Documents	Chatting	Coding	Games
Cosine similarity	90%	80%	87%	93%
Tanimoto coefficient	77%	87%	77%	94%

membrane keyboard when release-stage features or push–release-stage features are used. The reason might be that the structure of the mechanical keyboard is more complex than membrane keyboard, and the feedback when releasing a mechanical key contains much more information.

5.3.7.2 Experimental Results of Human-Computer Operation Recognition

To evaluate the performance of human-computer operation recognition, we collected data of 15 users with different input modes, i.e., fingering input and single-finger input [113]. If a user is used to fingering input, each key is usually pressed by the same finger and intensity. On the other hand, if a user is used to single-finger input, keystrokes are mostly performed by the forefinger and the input rate and intensity are distinct for different keys. For more information about the experimental setup, please refer to the article [113].

We first obtained segments of human-computer operations based on the feature vector similarity between adjacent segments, using the cosine similarity and the Tanimoto coefficient. With the data of four human-computer operations from single-finger input and fingering input, we tested the segmentation performance of the two metrics. The results are illustrated in Table 5.5, where 200 segments are used for each human-computer operation. Overall, the cosine similarity outperforms the Tanimoto coefficient.

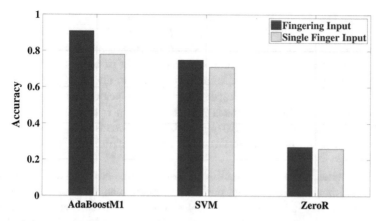

Fig. 5.20 Performance of different models for human-computer operation recognition

Furthermore, we recognized human-computer operations by applying three different classification models, including J48-based AdaBoost algorithm, ZeroR, and SVM. According to Fig. 5.20, we find that the AdaBoost model produces the highest recognition accuracy for both input modes, where the corresponding accuracies are 91% and 78%, respectively.

5.4 Swimmer Localization Based on Smartphone[4]

5.4.1 Introduction

Swimming pools are one of the major venues where accidental drowning occurs, with young children especially vulnerable [114]. Such tragic accidents can potentially be minimized, as swimming pools usually have trained lifeguards on duty to rescue victims, provided that they can get to them in time. This has prompted academic researchers [115, 116], as well as commercial companies [117], to propose systems to help identify potential drowning incidents on time.

Another equally important problem, apart from determining who are potential drowning victims, is to determine where the victims are exactly located. Lifeguards have a short period of time to reach a potential drowning victim, and an accurate method to pinpoint exactly where a swimmer is located in a pool is invaluable. This is, in essence, an entity localization problem which is actively studied in the research

---

[4]Part of this section is based on a previous work: D. Xiao, Z. Yu, F. Yi, L. Wang, C. Tan & Guo, B. (2016). SmartSwim: An Infrastructure-Free Swimmer Localization System Based on Smartphone Sensors. In: Chang C., Chiari L., Cao Y., Jin H., Mokhtari M., Aloulou H. (eds) Inclusive Smart Cities and Digital Health. ICOST 2016. Lecture Notes in Computer Science, vol 9677. Springer, pp. 222–234.

community [118–122]. Most existing localization techniques can be classified as either infrastructure-based techniques or device-based techniques.

Infrastructure-based localization relies on deploying infrastructure such as Wi-Fi access points to use the propagation of radio signals to estimate distance. Such techniques are accurate, but require the installation of additional infrastructure in swimming pools. However, existing device-based localization techniques cannot be applied directly to locate swimmers. The main reason is that, unlike walking or running, the swimmer may use different styles (e.g., breaststroke, freestyle) during swimming, which will cause different distances and different locations of the swimmer.

In this chapter, we propose SmartSwim, a smartphone-based system to accurately locate a swimmer in swimming pool. The smartphone running SmartSwim is encased in a waterproof carrier strapped onto the person while swimming. SmartSwim first learns the patterns for different styles of swimming (e.g., breaststroke, freestyle) by collecting the accelerometer and gyroscope values while people are swimming. It then automatically classifies the swim activity into different styles by using a sliding window technique and lightweight supervised learning algorithm. After identifying the specific style, SmartSwim then estimates the distance by calculating the location of a swimmer using the swimmer's original position, thus providing an accurate position of each swimmer in real time.

5.4.2   Related Work

Entity Localization: Infrastructure-based techniques use the Wi-Fi signal to determine the location of an entity [118, 119]. Closer to our work are smartphone-based techniques. Hsu et al. [120] designed an indoor localization system using the accelerometer and gyroscope sensors in a smartphone. Qian et al. [121] proposed an improved method using the inertial sensors with pedestrian dead reckoning (PDR) to determine the relative location change of a pedestrian.

Motion Detection with Wearable Devices: Wearable devices can be attached on user's wrist, knee, arm, and other body parts, which can detect the user's motion more accurately. Hassan et al. [123] proposed a wearable system that detects the user's running style using force-sensitive resistors in the insole of a running shoe and uses electric muscle stimulation (EMS) as a real-time feedback channel to intuitively assist the runner. Chun et al. [124] designed an instrumented necklace that captures head and jawbone movements for detecting eating episodes. However, such devices are not very prevalent and may be obtrusive. In specific, Seuter et al. [125] identified and quantified the impact of device interaction on the running activity.

Smartphone as a Sensor: Sports-related applications based on smartphone sensing include running [126] golf [127], and snowboarding [128]. In modern equestrian sports, Echterhoff et al. classified horse gait types and detected canter strides and jumps performed by a horse-rider pair using a smartphone attached to the horse's saddle [129]. Similar to our work, Marshall et al. [130] proposed a system to use the smartphone to track swim coaching, allowing swimmers to access timely feedback and improve their swimming skills. However, the system cannot give the locations of potential drowning victims.

Swim Tracking and Analyzing: There have been several studies that used nonsmartphone hardware to perform swim monitoring. Bächlin et al. [131] proposed a wearable assistant for swimmers, which consists of acceleration sensors with microcontrollers and feedback interface modules that a swimmer wears during swimming. The study of Siitola et al. [132] showed that tracking can be done with high accuracy using a simple method that is fast to calculate with a really low sampling frequency.

In general, the system we proposed aims at swim tracking and swimmer locating by leveraging on the inherent sensors on smartphones. Compared to prior research, our work mainly focuses on swimming status classification, swimming length estimation, and swimmer localization.

5.4.3 System Architecture

In this section, we introduce the proposed framework for supporting swimming assistant applications as illustrated in Fig. 5.21. The system consists of three components: data capture, swimming behavior recognition, and swimming localization. We first collect data from the built-in sensors in smartphone worn by a swimmer. Then, the data is preprocessed and features are extracted for swimming behavior recognition. Finally, the swimming localization is performed based on the recognized swimming behavior, stroke counts, and estimated depth.

Data capture: We used the accelerometer, gyroscope, and barometer sensors in a smartphone to capture the necessary data. Since a typical smartphone is not water resistant, we encased the phone in a waterproof pouch which is strapped onto the swimmer's lower back. Figure 5.22 shows the pouch and how it is worn by a swimmer.

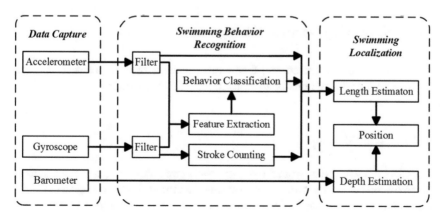

Fig. 5.21 SmartSwim system architecture

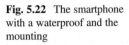

Fig. 5.22 The smartphone with a waterproof and the mounting

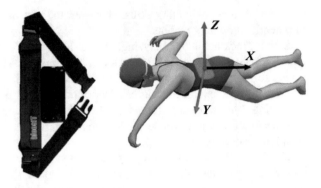

Swimming behavior recognition: This module uses the collected sensor data to determine the swimming style. It first preprocesses the raw data using a combination of moving average filter to identify outliers and a sliding window technique to sample the data. After processing, the component extracts several features and applies supervised learning models for classification. The swimming behavior recognition component is able to recognize three main swimming styles, i.e., breaststroke, freestyle, and backstroke, in addition to walking and turning.

Swimming localization: This module determines the actual real-time location of a swimmer. It consists of three parts: stroke counting, length estimation, and depth estimation. For stroke counting, we used the gyroscope data to detect each stroke. Then, a moving length estimation algorithm inspired by [133] is used to estimate distance. Unlike [133], however, our proposed system can estimate five types (breaststroke, freestyle, backstroke, walking, and turning) of moving length. Finally, depth estimation is accomplished using barometer as an indicator. The swimmer's position is calculated by combining these factors with the swimmer's starting location.

5.4.4   Swimming Behavior Recognition

We first filter the raw sensor, and then extract features from the cleaned data, and finally recognize the swimming style using the extracted features.

5.4.4.1   Data Filtering

We collect the sensor data from the smartphone's three-axis accelerometer and three-axis gyroscope. Another extra m-axis sensing data for each sensor is then calculated from the root-mean-square value of three-axis data individually as follows:

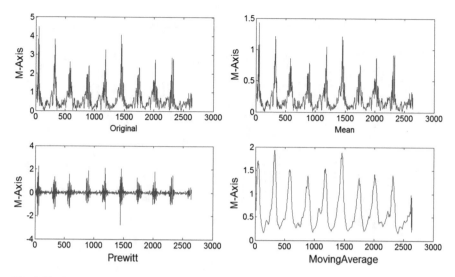

Fig. 5.23 Comparison of different filtering methods

$$m_s = \left(x_s^2 + y_s^2 + z_s^2\right)^{\frac{1}{2}}, \tag{5.11}$$

where s represents accelerometer or gyroscope sensor, and x, y, and z represent the three-axis data, respectively. Thus, we have totally eight dimensions, which are x, y, z, and m-axis data for accelerometer and gyroscope, respectively.

However, each dimension of the sensing data consists of outliers which distort the patterns in the data. We compared three different filtering algorithms: moving average filter, mean filter, and Prewitt horizontal edge-emphasizing filter. Figure 5.23 illustrates the results of applying these three techniques on the original data. From the results, we can see that the moving average filter performs the best, as it removed the noise in the data effectively, and the filtered results are close to the real value. The equation that the moving average method used is as follows:

$$G_{xyz\text{filterd}}[i] = \frac{1}{M} \times \sum_{j=-(M-1)/2}^{(M-1)/2} G_{xyz}[i+j], \tag{5.12}$$

where G_{xyz} is the original data obtained from the sensors, M is the window width, and $G_{xyz\text{filterd}}$ is the filtered data.

As indicated in Eq. (5.12), the different width of the window M is directly related to the sampling rate of the data. We then use a slide window technique with 50-window width (frequency 100 Hz) to acquire the most ideal data. In practice, we chose about 2-s-long window with a slide of 0.5 s between two sequential windows (as shown in Fig. 5.24). Therefore, every two consecutive windows will have an overlap of 1.5 s and we classify the ongoing activity in each half second. The method has been widely used in existing studies and proved effective [132, 134, 135].

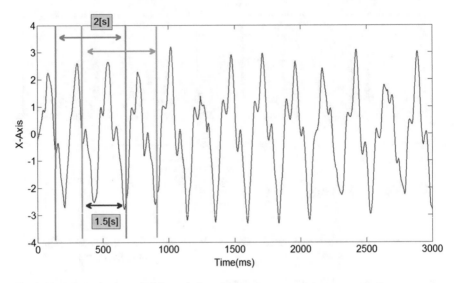

Fig. 5.24 Schematic view of sliding window

5.4.4.2   Feature Extraction

We classify swimming behavior into five types: breaststroke, freestyle, backstroke, walking, and turning. We found that different swimming behaviors result in different body movements or postures. For example, when a swimmer is performing the breaststroke, his/her waist arches upward with every stroke to lift his/her head above the water for air. This folding motion of the waist is less pronounced when he/she is utilizing freestyle. When swimming with the freestyle, the swimmer's body will rotate more with each stroke as he/she rotates his/her body sideways for air. Based on these observations, we extract three types of features which describe the posture of body activities as follows.

Body Folding Feature (BFF): A swimming style like breaststroke requires the swimmer's waist to be folded upwards every time he/she takes a breath, and thus BFF describes the level how a swimmer folds his/her waist upwards. This is obtained from the m-axis accelerometer records, which is defined as

$$\text{BFF} = \underset{0 \leq i \leq \text{WS}}{\text{Max}} \left(\text{Acceler}_i^m\right), \qquad (5.13)$$

where WS represents the window's size for data sampling and Acceler_i^m is the value in the corresponding window.

Body Rotation Feature (BRF): This feature describes the body rotation level while swimming. For example, when a swimmer performs an arm pull, his/her body is consequently rotated to keep balance. We use BRF as a factor to indicate the rotation of body and it is extracted from *m*-axis gyroscope records, which is defined as

$$BRF = \underset{0 \le i \le WS}{Mean} \left(Gyr_i^m\right) \tag{5.14}$$

where Gyr_i^m represents the m-axis data from gyroscope in each sampling window.

Breast Direction Feature (BDF): A swimmer's breast direction could be classified into two types in a swimming pool: vertical against the horizontal plane when he/she is standing and parallel with the horizontal plane while he/she is swimming. Thus, we use BDF to distinguish these two body postures which is extracted from different axis data of accelerometer as

$$BDF_{Ver} = \underset{0 \le i \le WS}{Max} \left(Acceler_i^x\right), \tag{5.15}$$

$$BDF_{Par} = \underset{0 \le i \le WS}{Mean} \left(Acceler_i^z\right). \tag{5.16}$$

where BDF_{Ver} and BDF_{Par} represent the vertical and parallel posture, respectively, $Acceler_i^x$ represents x-axis data from accelerometer, and $Acceler_i^z$ is z-axis data in each sampling window.

5.4.4.3 Behavior Recognition

We explored various supervised learning methods in our behavior recognition. One constraint is that the method should be efficiently implemented with the limited computational resources of the smartphone. We found that tree-based classifiers such as J48, LADTree, and RandomTree were suitable for our purpose. As concluded by Woohyeok et al. [135], the specific user model which uses one user's data for training and testing at the same time can achieve high accuracy. Thus, we build different models for different swimmers.

5.4.5 Swimmer Locating

In order to locate a swimmer's position in real time, we need to address three issues: swimming stroke counting, moving length estimation, and depth estimation. In this section, we introduce the methods solving these problems, respectively.

5.4.5.1 Swimming Stroke Counting

Based on our analysis of accelerometer, gyroscope, and barometer data, we found that m-axis gyroscope data can be used as an indicator for recognizing breaststroke cycle while x-axis gyroscope data can be used for backstroke and walking cycle. Figure 5.25 shows the triaxial gyroscope signal collected from a smartphone during

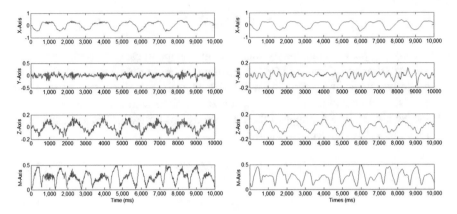

Fig. 5.25 Triaxial gyroscope data for breaststroke: raw and filtered data

breaststroke. We can observe that m-axis gyroscope data shows better periodicity than the other axis of data, and thus we adopt m-axis gyroscope data for cycle discovery on breaststroke.

Peak detection is a common method to discover cycles in time-series data, in which sliding window [136] and criteria algorithm [121] are widely used. Prior work usually adjusts the size of window to realize peak detection manually. But this results in poor peak recognition accuracy, as the algorithm is largely dependent on the size of the window. As shown in Fig. 5.26, one cycle of a stroke is detected by discovering the peaks during the swimming period. There are several features in each cycle of data (see Fig. 5.26, one cycle of a stroke ranges from A to B). On the one hand, it is obvious that during the period from A to B, the curve has to cross the threshold value δ_k from negative to positive twice, and the corresponding time slot is Δt_k. On the other hand, there must be two peaks within this time period and the corresponding amplitude is defined as ΔG_k.

However, as the cycle detection process is a real-time procedure, it is impossible to obtain the current Δt_k at time t_k, and the threshold for period Δt_k is undefined as well. In order to estimate a relative accurate value of the threshold δ_k for period Δt_k, we use an empirical formulation based on the information from last cycle which is already detected, and the equation is defined as follows:

$$
\delta_k = \begin{cases} \dfrac{1}{4 \times \mathrm{WS}} \displaystyle\int_0^{4 \times \mathrm{WS}} G_m(t)\mathrm{d}t, & k = 0 \\[3ex] \dfrac{1}{2 \times \Delta t_{(k-1)}} \displaystyle\int_{t_{(k-1)}}^{t_{(k-1)}+2 \times \Delta t_{(k-1)}} G_m(t)\mathrm{d}t, & k > 0 \end{cases} \tag{5.17}
$$

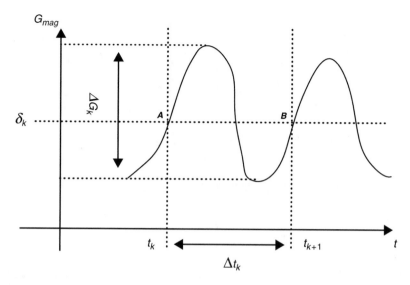

Fig. 5.26 One cycle of a stroke

where WS represents the window's size; thus, for the first time to estimate δ_k, we use the first four window data of G_m. The following δ_k for time period Δt_k is then estimated employing the time slot at t_{k-1}, which is Δt_{k-1} in particular. Therefore, after δ_k is estimated, it is possible to detect points A and B, respectively, for cycle detection as well as stroke counting.

Although the above algorithm with an appropriate threshold δ_k is able to estimate the cycle time, a cycle verification mechanism must be introduced. The reason is that there still exist small disturbances which would influence the accuracy for cycle detection.

Intuitively, there is little difference between two consecutive swim strokes as people swim regularly, which means Δt_k fluctuates around Δt_{k-1} and ΔG_k also fluctuates around ΔG_{k-1}. Thus, it is unacceptable when Δt_k or ΔG_k is much shorter than Δt_{k-1} or ΔG_{k-1}, respectively. Therefore, range feature (RF) mechanism is then applied to distinguish the reasonable Δt_k and ΔG_k using history information. Specifically, we define min $\Delta t_k^{\min}$, $\Delta t_k^{\max}$, $\Delta G_k^{\min}$, and $\Delta G_k^{\max}$ as follows:

$$\Delta t_k^{\min} = \alpha \times \frac{\left(\Delta t_{(k-1)}^{\min} + \Delta t_{(k-1)}\right)}{2}, \tag{5.18}$$

$$\Delta t_k^{\max} = \beta \times \frac{\left(\Delta t_{(k-1)}^{\max} + \Delta t_{(k-1)}\right)}{2}, \tag{5.19}$$

$$\Delta G_k^{\min} = \alpha \times \frac{\left(\Delta G_{(k-1)}^{\min} + \Delta G_{(k-1)}\right)}{2}, \tag{5.20}$$

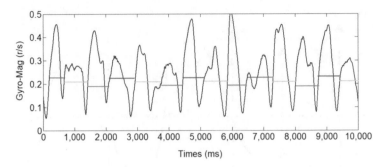

Fig. 5.27 Results of stroke detection for breaststroke

$$\Delta G_k^{\text{max}} = \beta \times \frac{\left(\Delta G_{(k-1)}^{\text{max}} + \Delta G_{(k-1)}\right)}{2}, \tag{5.21}$$

where Δt_k^{min} and Δt_k^{max} represent the upper and lower bounds for a candidate Δt_k, and ΔG_k^{min} and ΔG_k^{max} are the upper and lower bounds for a candidate ΔG_k. Two adjustment factors α and β are considered in practice which are empirically set to be 2/3 and 4/3, respectively. As for different swim style, we initiate Δt_k^{min} and Δt_k^{max} with different fixed values manually and the ΔG_0 is calculated within the first four windows. Then, if any Δt_k is larger than the corresponding Δt_k^{min} and smaller than Δt_k^{max}, as well as the value of ΔG_k is between ΔG_k^{min} and ΔG_k^{max}, respectively, we can say that a swim stroke is discovered.

Figure 5.27 illustrates the results of stroke detection for breaststroke with the above algorithm, in which nine strokes were detected. Once the stroke is discovered, it is easy to count the number of strokes; thus, each stroke length becomes the significant factor in moving length estimation.

5.4.5.2   Moving Length Estimation

Our moving length estimation method is inspired by prior work on localization of walking subjects [133]. We present a moving distance empirical formulation, which is learned from the observations of the relationship between step length and other factors during walking.

First, we calculate M_k, which is the moving distance, defined as

$$M_k = \sigma \times \frac{\sqrt[3.5]{\sum_{i=1}^{i=1} \frac{|a_i|}{N}} \times \sqrt[6]{a_{\text{peak,diff}}} \times \sqrt[7]{H}}{\sqrt[8]{\Delta t_k}} \tag{5.22}$$

where σ is an empirical coefficient with different fixed values in different strokes, a_i is the vertical acceleration, $a_{\text{peak, diff}}$ represents the difference between the maximum and minimum of vertical acceleration during one stroke, H is the height of swimmer, and Δt_k is the time for a cycle calculated by swimming stroke counting method. The empirical coefficient can be customized through adjusting the value of σ. Assuming that M_{real} is the ground truth of swimmer's moving distance, $M_{estimate}$ is the estimated value, and the next moving distance M_{new} is calculated as follows:

$$\sigma_{new} = \sigma_{old}^{\frac{M_{real}}{M_{estimate}}}. \tag{5.23}$$

5.4.5.3 Depth Estimation

The barometer sensor is commonly integrated in modern smartphones, and the data from this sensor can be translated to height [137]. We found that the barometer is capable to measure height change under water with low sensing error. For instance, the change in the ambient pressure of 1 hPa requires the change in depth of only 0.01 m under water; this change can be achieved by altering the height of 7.9 m in the air [138]. Using the above three methods, we can locate the user as

$$\begin{cases} X_{k+1} = X_k + C_k \times M_k \\ H_{k+1} = H_k + \Delta H \end{cases} \tag{5.24}$$

where X_k is the location on time k, X_0 is obtained from the user's original position, C_k is the stroke count, M_k is the average moving distance during the C_k strokes, H_k is the depth in the swimming pool of time k, and ΔH is calculated by using the barometer data.

5.4.6 Performance Evaluation

To evaluate the developed method, we recruited five college students to collect data for experiments. For more details about the experimental setup, please refer to the article [139]. Statistics of the obtained dataset is shown in Table 5.6.

5.4.6.1 Results of Swimming Behavior Classification

To demonstrate the effectiveness of the extracted features, we present the feature space of different feature pairs with different swimming behaviors in Figs. 5.34 and 5.35. According to Fig. 5.28, we can find that different magnitudes of value on BDF_{Ver} or BFF lead to different swimming behaviors. Specifically, when the value of BDF_{Ver} is lower than -5, a walking behavior can be identified (in red dots) and a backstroke can be detected (in pink stars) when the value of BFF is larger than 10.

Table 5.6 Swimming data

Swimmer ID	Swimming status				Total (m)
	Breaststroke	Freestyle	Backstroke	Walking	
1	40	65	70	50	225
2	30	35	55	50	170
3	50	70	60	50	230
4	28	45	60	50	183
5	32	55	70	50	207

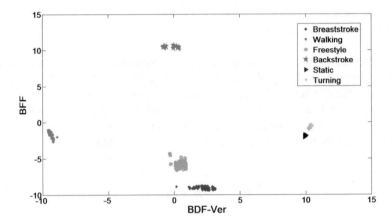

Fig. 5.28 Feature space of BDF_{Ver}-BFF

Similarly, according to Fig. 5.29, different types of swimming behaviors can also be differentiated while using the other two features BDF_{Par} and BRF. Therefore, it is possible to accurately classify different swimming behaviors by fusing the extracted features.

We summarize the ability of different features to distinguish different swimming behaviors in Table 5.7.

Accordingly, we can find that the four extracted features have different abilities for classifying different swimming behaviors. For example, BDF_{Ver} is effective for indicating walking, BDF_{Par} has the ability to distinguish backstroke, BFF can be used to differentiate breaststroke from backstroke, while BRF performs well in recognizing freestyle.

Based on these features, we trained three tree-based classification models: J48, LADTree, and RandomTree. The average accuracy for swimming behavior classification is 99.5%. Specifically, RandomTree outperforms the other two models with an accuracy of 99.75%.

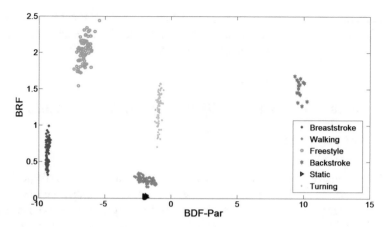

Fig. 5.29 Feature space of BDF$_{Par}$-BRF

Table 5.7 Effectiveness of different features

Feature	Possible classification
BDF$_{Ver}$	Walking-(breaststroke-freestyle-backstroke)-(static-turning)
BDF$_{Par}$	(Breaststroke-freestyle)-(static-walking-turning)-backstroke
BFF	Breaststroke-freestyle-(static-walking-turning)-backstroke
BRF	Freestyle-(backstroke-turning)-(breaststroke-walking)-static

Table 5.8 The accuracy of stroke counting and moving length estimation

	Accuracy of stroke counting			Accuracy of moving length estimation		
Status	Real	Experiment	Error rate (%)	Real (m)	Experiment (m)	Error rate (%)
Breaststroke	50	47.5	5.00	36	34.2	5.00
Freestyle	60	57.5	4.17	54	57.8	7.03
Backstroke	50	48	4.00	63	67.5	7.14
Walking	100	96	4.00	50	47.3	5.40

5.4.6.2 Results of Stroke Counting and Moving Length Estimation

To test the performance of the proposed stroke counting algorithm and the moving length estimation algorithm, we calculated the average error rate of these two algorithms for all the swimmers; the results are demonstrated in Table 5.8.

As we can see, the overall error rate of stroke counting is lesser than 5%. Specifically, the counting accuracies of breaststroke and walking are a little bit higher and the corresponding error rates are less than 4%. Furthermore, the error rate of moving length estimation is around 7% or less, and the estimation accuracy of breaststroke is the highest with an error rate of 5%.

5.5   Human Identity Recognition Based on Wi-Fi Signals[5]

5.5.1   Introduction

Generally, human identification is based on one or more intrinsic physiological [140–143] or behavioral [144–146] distinctions, which is related to either the shape of the body (e.g., fingerprint, iris, palm print, face characters) or certain behavior patterns of a person (e.g., typing, gait, voice rhythms). It plays a significant role in the area of pervasive computing and human-computer interaction. Currently, fingerprint- [141], iris- [143], and vein-based methods [147] have been successfully employed in automatic human identification systems. However, these systems require the user to be close to the sensing device for accurate identification. Researchers also make numerous attempts to develop methods for behavioral biometrics (mainly gait analysis) using cameras, radars, or wearable sensors [146]. However, the vision-based approaches only work with line-of-sight coverage and rich-lighting environments, which also cause privacy concerns. The low-cost 60 GHz radar solutions can only offer an operation range of tens of centimeters [148], and the devices are not widely deployed in our daily living environments. Finally, wearable sensor-based approaches require people to wear some extra sensors.

Wi-Fi techniques have been widely used in our daily life. Due to its popularity and low deployment cost, numerous studies have devoted to pervasive sensing using Wi-Fi devices, such as indoor localization [149] and gesture recognition [150]. These studies are mainly based on received signal strength indication (RSSI), i.e., the coarse-grained signal strength information. Quite recently, channel state information (CSI), i.e., fine-grained information regarding Wi-Fi communication, becomes available. Specifically, CSI describes how the signal propagates from the transmitter to the receiver and reflects the combined effects of the surrounding objects (e.g., scattering, fading, and power decay with distance). There are many subcarriers in CSI, each of which contains the information of attenuation and phase shift. Therefore, CSI contains rich information and is more sensitive to environmental variances caused by moving objects [151]. Several notable studies on pervasive

[5]Part of this section is based on a previous work: T. Xin, B. Guo, Z. Wang, M. Li, Z. Yu and X. Zhou, "FreeSense: Indoor Human Identification with Wi-Fi Signals," 2016 IEEE Global Communications Conference (GLOBECOM), Washington, DC, 2016, pp. 1–7. DOI: https://doi.org/10.1109/GLOCOM.2016.7841847

sensing have been conducted using CSI, such as high-accuracy human localization [152], human activity recognition (gesture, fall detection, etc.) [153–159], and crowd counting [151]. In this chapter, we present how to realize human identification based on Wi-Fi CSI.

Due to the difference of body shapes and motion patterns, each person can have specific influence patterns on surrounding Wi-Fi signals while he/she moves indoors, generating a unique pattern on the CSI time series of the Wi-Fi device. According to such an intuition, we propose a novel approach called FreeSense, which can identify human indoors based on CSI-enriched Wi-Fi devices. It is supposed to work in home environments, which usually have 2–6 family members (the 2014 CPS ASEC research shows that more than 98% American family have less than 6 members) and deliver personalized services when recognizing the identity of a family member. There are two technical challenges faced by such a system, as presented below.

Identification Condition Settlement and CSI Time-Series Segmentation. When a person moves around in the house, the influence level over multipath communication of Wi-Fi signals can be different when people walk across different paths. It is easy to understand that the identification performance can be raised at high human-Wi-Fi-signal influence level, where the characterizing features are more distinguishable. In other words, it is difficult to identify human using the waveform caused by arbitrary movements. Therefore, we should find a proper functional window to maximize the identification performance. The waveform obtained within the functional window will be used for human identification.

CSI Feature Extraction for Human Identification. Different people have different influence patterns to Wi-Fi signals while moving around, regarding their body shapes and motion patterns. To conduct exact human identification over the extracted line-of-sight (LOS) waveform, we should utilize proper features to characterize the influences. Ordinary features such as the rate of change, signal energy, and maximum peak power are not effective, as they all have similar values to different people. The extraction of useful CSI features, however, becomes another challenge to be studied in our work.

To address these issues, FreeSense makes the following contributions:

- We propose a novel approach called FreeSense to identify human indoors based on Wi-Fi CSI signals, which is nonintrusive and privacy preserving compared to existing methods.
- We identify the line-of-sight path-crossing moments as the functional window for enhanced human identification. Moreover, we put forward an algorithm to segment the CSI time series for the extraction of LOS waveforms within the functional window. It is based on the observation that the LOS waveform caused by human walking across LOS path shows a typical increasing and decreasing trend in rates of change in CSI time series.
- We use a model to extract the feature of LOS waveform, which consists of principal component analysis (PCA) and discrete wavelet transform (DWT). The recognition is based on the difference of personal movement influence to

Wi-Fi signals. We first use PCA to get the principal component of CSI time series, and then adopt DWT to compress the waveform and extract the shape feature of LOS waveform.

- We deployed a prototype system in a 6 m∗5 m smart home environment and recruited nine users for evaluation. The results show that we can achieve 92.6% detection rate for CSI functional window segmentation, and achieve 88.9–94.5% accuracy when the candidate user sets changes from 6 to 2, which is effective in domestic environments.

5.5.2 Related Work

5.5.2.1 Human Identification

Researchers have made numerous attempts to develop methods for human identification, which can be grouped into two categories: physiological feature-based approaches and behavioral feature-based approaches.

Physiological features. Fingerprint [141], iris [143], and vein authentications [147] have been successfully employed in automatic human identification systems. Furthermore, Duta [140] used the hand shape to distinguish human. Chellappa et al. [160] proposed a method to recognize human based on face recognition. However, these systems require the subject to be close to the sensor for accurate identification.

Behavioral features. There are also some studies about behavioral biometrics, especially in gait analysis. Since human walking motion results in a self-sustaining and dynamic rhythm owing to the integrated signals generated from the spinal cord and sensory feedback, it contains unique characteristics of each individual. Little and Boyd [145] developed a model-free description of instantaneous motion and used it to recognize individuals by their gait. The work is based on camera sensing, which requires line of sight and enough lighting, and it also causes privacy issues. Nickel et al. [146] used HMM to recognize human gait based on accelerometer data, and it requires users to wear relevant sensors. We use Wi-Fi signal to recognize human, which can provide better coverage and are device free.

5.5.2.2 CSI-Based Motion Detection

CSI values can be obtained from COTS Wi-Fi network interface cards (NICs) (such as Intel 5300 [161] and Atheros 9390 [162]). It is used for human activity recognition and indoor localization. Han et al. proposed WiFall, which can detect falls of human in indoor environments by analyzing CSI values [154]. E-eyes proposed by Wang et al. exploits CSI for recognizing household activities such as washing dishes and taking a shower [157]. Nandakumar et al. leveraged the CSI and RSS information offered by off-the-shelf Wi-Fi devices to classify four arm gestures—push, pull, lever, and punch [155]. Wang et al. proposed WiHear, which uses CSI to recognize the shape of mouth while human speaking [156]. Ali et al. proposed WiKey, which can recognize keystrokes by using CSI [153]. For indoor localization, Li et al. [163] proposed MaTrack to detect the subtle reflection signals from human body and further differentiate these signals from those reflected from static objects (furniture, walls, etc.) to identify the target's angle for localization. They developed IndoTrack [164] to estimate the absolute trajectory of human. Moreover, Zhang et al. [165, 166] introduced the Fresnel zone model to the indoor environment, using the model to depict RF signal propagation properties and develop techniques for indoor respiration detection [167] and direction estimation [168], based on commodity Wi-Fi devices. Zhang et al. [169] also introduced the Fresnel diffraction model for human respiration monitoring within the first Fresnel zone (FFZ), which provides a general theoretical foundation for fine-grained human and object movement sensing. Different from these studies, our work, however, concentrates on human identification, which is not investigated in existing CSI-based studies.

5.5.3 Problem Analysis and System Framework

5.5.3.1 Problem Analysis

Identification Condition Settlement: According to the above analysis, to identify human identity we need to select an appropriate identification condition. Although the walking routes of people are difficult to control, we can easily make the users walk across the LoS path by properly deploying the devices. Meanwhile, based on observations from a large amount of data, we find that the Wi-Fi signal has a more notable fluctuation when people walk across the LoS path. Figure 5.30 shows the amplitudes of CSI time series of five different subcarriers of a transmitter-receiver antenna pair, where a person repeatedly walks across the LoS path. Furthermore, the signal of LoS path is more stable than N-LoS path. Thus, we propose to identify human by analyzing waveforms produced when the user walks across the LoS path. Specifically, we call this waveform the line-of-sight (LoS) waveform.

Feature Extraction: Different people have different influence patterns to Wi-Fi signals while moving around, regarding their body shapes and motion patterns. To conduct exact human identification over the extracted LoS waveform, we should utilize proper features to characterize the influences. Ordinary features such as the rate of change, signal energy, and maximum peak power are not effective, as they all have similar values to different people. According to experimental observations, we find that the unique human influence on Wi-Fi signal may reflect on both time and frequency domain features. Therefore, we use the shapes of extracted LoS waveform as features, which retains both time and frequency domain information of the waveform.

Each transmit-receive (TX-RX) antenna pair of a transmitter and receiver has 30 subcarriers. Let MTx and MRx represent the number of transmitting and receiving antennas. There are $30 * MTx * MRx$ CSI streams in a time series of CSI values.

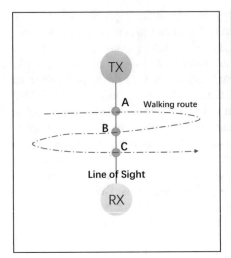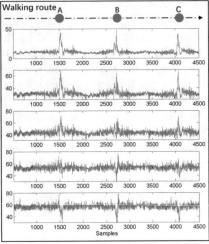

Fig. 5.30 The waveform of walking: the route of walking (left) and the amplitudes of CSI time series (right)

Directly using the 30 ∗ MTx ∗ MRx CSI streams will lead to high computational cost. We thus need to compress the data. As Fig. 5.30 shows, all subcarriers show correlated variations in their time series. Thus, we use principal component analysis (PCA) to reduce the dimensionality of the CSI time series. Subsequently, we also apply discrete wavelet transform (DWT) to compress the extracted LoS waveform, which can preserve most of the time and frequency domain information.

Classification Method: The starting time and ending time of LOS waveforms determined by the extraction algorithm are not exact, which makes the midpoints of extracted line-of-sight waveforms rarely align with each other. Meanwhile, the lengths of different waveforms caused by different human walking are also different. Thus, two waveforms cannot be compared using standard measures like correlation coefficient or Euclidean distance. To address this issue, we adopt the dynamic time warping (DTW) technique, which can find the minimum distance alignment between two waveforms of different lengths.

5.5.3.2 System Framework

As shown in Fig. 5.31, our system consists of three parts: data collection and noise removing, feature extraction, and human identification.

Data Collection and Noise Removing: We first record the CSI value with COTS Wi-Fi network interface cards (NICS), while people walk across the LOS path. Afterwards, we choose a suitable filter to remove noise.

Fig. 5.31 The FreeSense framework

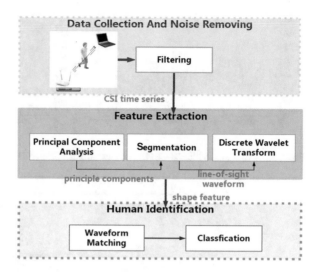

Feature Extraction: We extract shape features of the LOS waveform from the CSI time series, which can be used to recognize human identity.

Human Identification: Based on the extracted features, a k-nearest neighbor (KNN) classifier is used for human identification.

5.5.4   Detailed Design of Human Identification

5.5.4.1   Noise Removing

The CSI values provided by commodity Wi-Fi NICS are inherently noisy. Thereby, we should remove the noise at first. We adopt the Butterworth IIR filter here which does not significantly distort the phase information in the signal and has a maximally flat amplitude response in the passband. According to experiments, we find that the frequency f of the variations in CSI time series caused by human walking is around 10 Hz. As the sample rate of CSI data is 1000 sample/s, the cutoff frequency w_c of Butterworth filter is calculated as

$$w_c = \frac{2\pi * f}{F_s} \approx 0.06\,\mathrm{rad/s} \tag{5.25}$$

The filtering result is shown in Fig. 5.32.

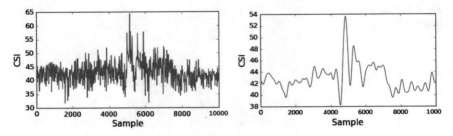

Fig. 5.32 Original and filtered CSI time series: original time series (left), filtered time series (right)

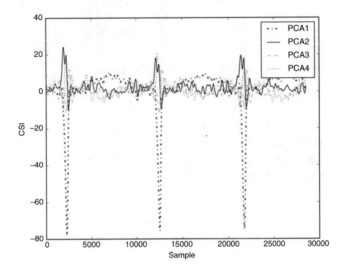

Fig. 5.33 PCA of CSI stream

5.5.4.2   Feature Extraction

To extract features from the CSI time series, we perform PCA in each TX-RX antenna pair and then segment the CSI time series to identify the starting point and ending point of each line-of-sight waveform. Finally, we adopt DWT to compress these waveforms.

Principal Component Analysis: For each TX-RX antenna pair, we first normalize the data such that every CSI stream has zero mean and unit variance, and then perform PCA on normalized CSI streams to get p principal components. As a result, we have $p * MTx * MRx$ waveform shapes for each human. Based on the observation that only the top four principal components contain significant variations in CSI values caused by human walking, we set $p = 4$. Figure 5.33 shows the top four principal components of one TX-RX antenna pair. In practice, we find that the PCA may lead to different principal components or different ordering of principal components in waveforms of different walking of the same person, because the result of

PCA is based solely on the value of their variance. To solve it, we also save the projection matrix of each training data when performing PCA. And when comparing the test data with a training data, we use the same projection matrix to transform it.

CSI time-series segmentation: To obtain the line-of-sight waveform from CSI time series, we need to identify the start point and end point of the waveform. As shown in Fig. 5.33, the waveforms caused by human walking across the LoS path have a typical increasing and decreasing trend in rates of change in the amplitudes of CSI time series. Based on this characteristic, the WiKey system [153] proposed an algorithm based on mean absolute deviation (MAD) to segment the CSI time series. However, it will cause high error rate if we directly use it in our work, due to the interference waveform caused by human walking across other paths. Instead, while we also design the segmentation algorithm based on this characteristic, a sliding window approach is adopted to search the segmentation point, which has an increasing and decreasing trend before or after it. Moreover, to reduce the error rate, we add more qualifications. The segmentation algorithm consists of the following four steps:

First, for every jth point of principal components $1 \leq k \leq p$, the algorithm calculates the mean absolute deviation (MAD) of each window of size w before and after it as follows:

$$
\begin{aligned}
\mathrm{MAD}^j_{\mathrm{before}}[k] &= \frac{\sum_{i=j}^{j+w} |\, Y(i) - \overline{Y}(\,j-w:j)\,|}{w} \\
\mathrm{MAD}^j_{\mathrm{after}}[k] &= \frac{\sum_{i=j}^{j+w} |\, Y(j) - \overline{Y}(\,j:j+w)\,|}{w}
\end{aligned}
\tag{5.26}
$$

where $\overline{Y}(\,j-w:j)$ represents the means of the CSI stream in the window before jth point and $\overline{Y}(\,j:j+w)$ represents the means of the CSI stream in the window after jth point.

Second, we calculate the sum of MAD of all p principal components in jth point as follows:

$$
\begin{aligned}
\mathrm{MAD}^j_{\mathrm{before}} &= \sum_{k=1}^{p} \mathrm{MAD}^j_{\mathrm{before}}[k] \\
\mathrm{MAD}^j_{\mathrm{after}} &= \sum_{k=1}^{p} \mathrm{MAD}^j_{\mathrm{after}}[k]
\end{aligned}
\tag{5.27}
$$

Third, the algorithm determines thresholds T_1 and T_2, $T_1 > T_2$. For every start point j_{begin} and end point j_{end}, it should satisfy the following condition:

$$
\begin{aligned}
\mathrm{MAD}^{j_{\mathrm{begin}}}_{\mathrm{before}} &\leq T_2, \mathrm{MAD}^{j_{\mathrm{begin}}}_{\mathrm{after}} \geq T_1 \\
\mathrm{MAD}^{j_{\mathrm{end}}}_{\mathrm{after}} &\leq T_2, \mathrm{MAD}^{j_{\mathrm{end}}}_{\mathrm{before}} \geq T_1
\end{aligned}
\tag{5.28}
$$

Finally, we have got a set of start points and end points, but not all of the points are qualified ones. Thus, we introduce the following limitations: (1) we record the end point only if we have got a start point, and (2) $timelen_1 \leq j_{end} - j_{begin} \leq timelen_2$, where $timelen_1$ and $timelen_2$ are two thresholds that are empirically determined. So far, we get a set of point pairs $<j_{begin}, j_{end}>$, based on which we can obtain the line-of-sight waveforms.

Discrete Wavelet Transform: We use DWT to compress every line-of-sight waveform, which can preserve most of the time and frequency domain information. Specifically, according to experimental results, we choose the Daubechies D4 wavelet and use approximation coefficients to represent the shape feature of waveforms.

5.5.4.3 Classification

After obtaining the shape features, we need a comparison metric that provides an effective measure of similarity between shape features of two LoS waveforms. However, the starting time and ending time of LoS waveforms determined by the extraction algorithm are not exact, which makes the midpoints of extracted line-of-sight waveforms rarely align with each other. Meanwhile, the lengths of different waveforms caused by different human walking are also different. Thus, two waveforms cannot be compared using standard measures like correlation coefficient or Euclidean distance. To address this issue, we adopt the dynamic time warping (DTW) technique, which can find the minimum distance alignment between two waveforms of different lengths.

We utilize k-nearest neighbor (KNN) classifier to classify a detected target. The KNN method does not have training stage and conducts classification tasks by first calculating the distance between the test sample and all training samples to obtain its nearest neighbors and then conducting KNN classification. It is popular because of its simple implementation and significant classification performance. Different from traditional KNN classifier which adopts Euclidean distance as the measurement criteria between samples, we use DTW distance as the comparison metric. Specifically, given shape features of two line-of-sight waveform series, X and Y, the distance between them is calculated as

$$\text{dis}(X, Y) = \sum_{m=1}^{N_T * N_R} \sum_{k=1}^{p} \text{DTW}_m(X_k, Y_k). \tag{5.29}$$

where p is the dimension of the line-of-sight waveform of each *TX-RX* antenna pair, and $M_T * M_R$ is the number of antenna pairs. The classifier searches for the majority class labels among k nearest neighbors based on the corresponding shape features using the DTW distance metric.

Table 5.9 Accuracy for segmentation

Method	WiKey	Our method
Detection ratio	91.7%	92.6%

Table 5.10 Error rate for segmentation

Method	WiKey	Our method
Error ratio	16.2%	6.5%

5.5.5   Performance Evaluation

To evaluate the proposed approach, we adopted a Lenovo X200 laptop with Intel Link 5300 Wi-Fi NIC as the receiver and a TP-Link TLWR1043ND Wi-Fi router as the transmitter. Specifically, the transmitter has two antennas and the receiver has three antennas, i.e., MT = 2 and MR = 3. We recruited nine subjects as volunteers to perform the experiment; each of the volunteers had provided 40 data samples. For more information about experimental setup, please refer to the article [170].

5.5.5.1   Line-of-Sight Waveform Detection Accuracy

To evaluate the performance of line-of-sight waveform detection, we adopted the WiKey algorithm as the baseline, and compared it with the proposed algorithm based on two metrics: detection ratio (DR) and error ratio (ER). The results are shown in Tables 5.9 and 5.10; accordingly we find that both of the algorithms achieved high detection ratios, while our algorithm had a lower error ratio.

5.5.5.2   Classification Accuracy

We used 20 samples from each person to train the classifier, and the rest samples for testing. The identification accuracy ranged from 94.5% to 75.5% while the number of users varied from 2 to 9. Specifically, the accuracy declines with the increase of users, as shown in Fig. 5.34. The reason is that as more people who have similar body shapes or motion patterns are involved in the system, the difficulty of identification would increase. It is worth to mention that as most of the family size ranges from 2 to 6, the identification accuracy (up to 88.9%) is effective and acceptable for family use in home environments.

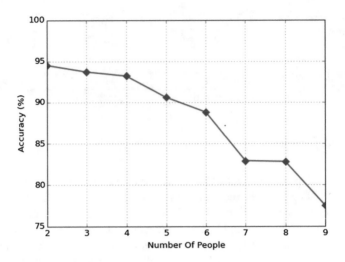

Fig. 5.34 Identification accuracy for 2–9 subjects

5.6 C-FMCW-Based Contactless Respiration Detection Using Acoustic Signals[6]

5.6.1 Introduction

Recent years have witnessed the surge of ubiquitous sensing technologies and their applications in vital sign monitoring [171, 172]. Respiration rate is one of the important vital signs, often used to assess and track human health condition. For instance, research studies show that a deep breath helps reduce blood pressure while irregular respiration patterns may indicate chronic stress, even progression of illness and decline in health [173–175]. Respiration monitoring in bed also helps track sleep quality and allows timely intervention in case of respiration abnormality such as obstructive or central sleep apnea-hypopnea which has been found quite common among the elderly [176–178]. In particular, chronic obstructive pulmonary disease (COPD) is the third most common cause of death for people aged 65 and above [179]. As such, respiration monitoring is essential to enable various ubiquitous healthcare applications. Since these systems often target patients or the elderly with chronic disease, a convenient and nonintrusive system is highly desirable.

Several respiration-monitoring systems have been proposed in clinical setting in early days such as thoracic impedance pneumography [180] and capnography [181];

---

[6]Part of this section is based on a previous work: T. Wang, D. Zhang, Y. Zheng, T. Gu, X. Zhou, B. Dorizzi. C-FMCW Based Contactless Respiration Detection Using Acoustic Signal. Proceedings of the ACM on Interactive, Mobile, Wearable and Ubiquitous Technologies table of contents archive, 1(4), December 2017, Article No. 170

however these systems are inconvenient and intrusive. For instance, these systems typically require users to attach/wear special devices such as nasal probes or chest bands, thus restricting body movements and causing inconvenience in their daily lives. Moreover, clinical respiration-monitoring systems typically require well-trained professionals to set up the systems and assist users to attach/wear respiration-monitoring devices properly. To minimize the discomfort brought by the invasive respiration systems, several attempts have been made for long-term respiration monitoring leveraging on wearable devices [182, 183]. While these systems are made more tolerable for the elderly, additional issues may arise such as user acceptance and usability [184].

Contactless sensing technologies explore a new possibility of monitoring vital signs without direct and physical contact to users. For instance, Penne et al. use a time-of-flight camera and apply advanced image processing algorithms to estimate human respiration rate in a home [185]. Unfortunately, such camera-based approaches require a subject to face the camera closely besides having privacy concerns and being affected by the lighting conditions. Kondo et al. deploy a laser sensor to measure the chest wall motion during respiration [186]. RF-based methods have also been widely studied, ranging from Doppler radar [187], UWB radar [188], and FMCW radar [189] to USRP (Universal Software Radio Peripheral)-based solutions [190, 191]. The basic idea of these methods is to detect the chest movement displacement during breathing. Though these methods are able to accurately detect human respiration, they incur prohibitive cost, preventing these systems from large-scale deployment in typical home settings. To discover cost-effective solutions, researchers have recently turned their attentions to widely available commodity WiFi or Zigbee devices. These works can be further classified into two subcategories: channel state information (CSI)-based method [192–196] and received signal strength (RSS)-based method [197–200]. Limited by the granularity of RSS, the RSS-based approach suffers from low resolution, and it is subjected to interference in the environment. The CSI-based method typically yields a higher resolution due to fine-grained channel information used. However, the study in [192] shows that the minute body movement may not generate detectable CSI changes. As such, the CSI-based approach typically calls for careful selection of device locations and subcarriers to achieve robustness.

In recent years, acoustic signal has been used to enable accurate affordable contactless reparation monitoring in home environments. The study in ref. 201 uses 40 kHz ultrasonic signal to detect human respiration by sensing the exhaled airflow. The Doppler effect introduced by the velocity of exhaled airflow can be used to detect respiration. While the idea of using acoustic signal to detect respiration is promising, these systems are vulnerable to ambient airflow, such as wind or even the airflow introduced by body movement around it. Frequency-modulated continuous wave (FMCW) has been first used to detect human respiration using acoustic signal by measuring the chest movement displacement during breathing [202]. It achieves about 0.7 cm distance estimation resolution and can detect the respiration with chest movement displacement of about 2 cm. However, the average respiration amplitude, i.e., the chest movement displacement, is only about 1 cm. Especially for Asian people whose body size is relatively thin, newborns, or the elderly, the chest movement displacement is usually within 1 cm [192], and this is beyond the detection capability of the FMCW-based systems [202].

Due to the fact that the distance estimation resolution of FMCW is restricted by the sweep bandwidth, it is difficult to achieve very high distance estimation resolution (i.e., detecting human respiration with chest movement displacement of less than 1 cm.) with narrow-band commodity acoustic device. In this chapter, we propose a correlation-based FMCW (C-FMCW) approach to reliably monitor respiration with commodity audio devices. Unlike traditional FMCW, C-FMCW estimates the round-trip propagation time of acoustic signals by discovering the maximum correlation between transmitted signal and received signal. For digital FMCW signal, the round-trip propagation time can be measured by detecting the number of samples corresponding to the maximum correlation. As such, the distance estimation resolution of C-FMCW is limited only by the acoustic signal sampling rate. With current prevailing audio systems of 48 kHz, our approach can achieve a ranging resolution of around 0.4 cm. The ranging resolution can be further improved if higher sampling rates (e.g., 96 kHz) are supported by audio systems. In C-FMCW, linear chirps facilitate the correlation detection, and also tolerate the frequency-selective fading of audio signals due to multipath effect.

Implementing such an audio-based respiration system however entails several practical challenges. First, the sampling frequency offset between commodity speakers and microphones can seriously affect the correlation-based detection. In practice, the sampling frequency offset squeezes or stretches audio signals such that the received signals and the transmitted signals cannot be aligned during the correlation process. Such an offset if not corrected properly may incur high ranging error. Second, in practice, even though we send a starting command to speaker and microphone simultaneously, the actual start time of the speaker is essentially different from that of the microphone. What is worse, the starting time difference varies every time we send a starting command to them, which will introduce intolerable absolute distance estimation errors. Third, due to multipath effect, weak reflected signals capturing user's respiration can be overwhelmed by stronger signals reflected from the static body parts of the user. It still remains elusive to extract the reflected signals capturing respiration.

We hence propose three novel techniques to address the above challenges. First, we propose an algorithm to compensate dynamically to the transmitted signal and counteract the signal shift caused by sampling frequency offset between speaker and microphone. Second, we co-locate speaker and microphone; thus, part of the transmitted signal will be directly received by the microphone without reflection (self-interference). We regard this part of received signal as a reference to estimate the absolute distance. In this way, we can counteract the starting time difference between speaker and microphone since the starting time difference remains the same for the signal without reflection and the signal reflected from abdomen. Third, based on the fact that respiration is almost periodic in nature, we extract periodic components by using autocorrelation to capture respiration while filtering out static signals.

The main contributions of this chapter can be summarized as follows:

- We propose a novel C-FMCW method for human respiration monitoring, which is able to achieve sub-centimeter ranging accuracy. C-FMCW adopts the correlation-based detection method to accurately measure the propagation time of audio signals with high-resolution audio samples.
- We propose several novel ideas to address practical issues, such as developing a compensation algorithm to dynamically add artificial shift to the transmitted signal to counteract the signal shift caused by sampling frequency offset between speaker and microphone; leveraging the received signal without reflection to eliminate the absolute distance estimation error introduced by the starting time difference between speaker and microphone; and leveraging the prior knowledge of respiration periodicity to enhance and extract weak periodic signals which can be otherwise overwhelmed by stronger reflection signals.
- We conduct extensive experiments to evaluate our system with 22 subjects. The experimental settings vary in different sleep postures, different rooms, and different scenarios. We also verify the system robustness with respect to different sensing distances, different scenarios, different respiration rates, different kinds of simulated apnea, and body movement. The results show that our system achieves a median error of 0.35 breaths/min.

5.6.2   Related Work

In this section, we briefly review the existing contactless respiration detection work, which can be roughly grouped into two categories:

5.6.2.1   Customized Device-Based Methods

Recently, customized device-based technologies ranging among laser [186], infrared [203], Doppler radar [187], UWB radar [188], FMCW radar [189], USRP [190, 191], and ultrasonic sensor [201] have been proposed to detect human respiration. The basic idea is to measure the chest movement displacement during respiration. Although these methods achieve accurate respiration detection, they typically require costly signal transceivers or customized radio front ends. As such, these methods are less appealing than the commodity device-based methods for large-scale deployment in home settings.

5.6.2.2   Commodity Device-Based Methods

In recent years, researchers have turned their attentions to commodity devices since they are readily available in home environments. The most widely studied methods are WiFi based and audio based.

WiFi-Based Methods: Ubibreathe [197] measures the received signal strength (RSS) to extract the respiration rate of a person in vicinity. RSS-based methods, however, require the subject in the line-of-sight path. WiSleep [194] and the studies in [192, 193, 195, 196] use the fine-grained channel state information (CSI) in commodity WiFi cards to detect respiration rate. Wang et al. [192] apply the Fresnel zone model to characterize the radio propagation in indoor environments and reveal the principle and limitations of CSI-based human respiration sensing with WiFi signals. According to their findings, CSI-based approaches cannot always reliably track human respiration and require careful selection of relative distance between human and WiFi devices.

Audio-Based Methods: ApneaApp [202] is a mobile application that allows smartphone users to monitor respiration rate. ApneaApp sends audio chirps and measures frequency shifts to monitor the breathing patterns so as to detect Apnea. The ranging resolution of ApneaApp is fundamentally limited by the narrow sweeping band of the device. As the ranging resolution of ApneaApp is only 0.7 cm, it can detect a deep breath with chest displacement of around 2 cm. Our system offers a more fine-grained ranging resolution of 0.4 cm within the 80 cm range. Moreover, the maximum ranging error is less than 0.2 cm when the distance between the acoustic devices and subject is from 40 cm to 60 cm, which can reliably detect a normal breath of young kids or elders who typically have a chest displacement of less than 1 cm. iSleep [204] uses microphone on a smartphone to detect the events that are closely related to sleep quality such as body movement and snore. However, iSleep cannot detect respiration. BeepBeep [205] detects the round-trip time of flight at the resolution of audio sampling rate and achieves high ranging accuracy between two smartphones. BeepBeep requires both smartphones to actively generate audio chirps, while our work detects the time of flight of reflected audio chirps. RunBuddy [206] measures the breathing pattern of a runner with Bluetooth headset and suggests a proper running rhythm. The headset is not intrusive for a runner but may cause inconvenience to a user during sleep.

In our work, we propose a novel acoustic respiration detection method leveraging commodity acoustic devices, which achieves a ranging resolution of around 0.4 cm. Based on the high ranging estimation resolution, we manage to detect human respiration with the chest movement displacement of less than 1 cm.

5.6.3   C-FMCW: A High-Resolution Distance Estimation Method

5.6.3.1   FMCW and Its Limitation

The naïve way to estimate distance d to an object using wireless signal is to transmit and receive a sharp pulse and compute the time delay τ, and then estimate the distance using their relationship $d = \tau \cdot v$, where v denotes the speed of wireless signal. However, sending a sharp pulse requires a large bandwidth. Instead, FMCW

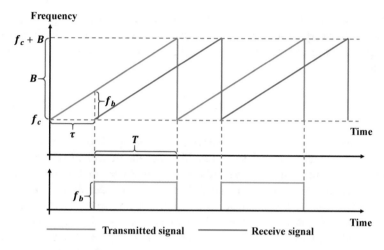

Fig. 5.35 Chirp signal in FMCW

[207] indirectly estimates the propagation delay based on the frequency shift of the chirp signal as follows.

As shown in Fig. 5.35, the red line denotes the transmitted signal, whose frequency linearly increases over time. The frequency of the transmitted signal at time t is given by $f(t) = f_c + \frac{Bt}{T}$, where f_c, B, and T denote carrier frequency, sweep bandwidth, and sweep period, respectively. The phase of the transmitted signal is calculated by integrating $f(t)$ over time,

$$u(t) = \int_0^t f(t')dt' = 2\pi\left(f_c t + B\frac{t^2}{2T}\right),\tag{5.30}$$

and the transmitted signal is then presented as $v_{tx}(t) = \cos(u(t))$. For simplicity, the amplitude is assumed to be 1. The transmitted signal is reflected by the target with the time delay τ. If amplitude attenuation is ignored, the received signal at time t can be represented as $v_{rx}(t) = \cos(u(t - \tau))$ (the blue line in Fig. 5.35). Consider that the transmitter and receiver are co-located. τ is calculated as

$$\tau = 2(R + vt)/C,\tag{5.31}$$

where R denotes the distance between transceiver and target, v denotes the moving speed of the target, and C denotes the propagation speed. The receiver multiplies the transmitted signal with the received signal as $v_m = v_{tx} \cdot v_{rx}$. By using $\cos A \cdot \cos B = (\cos(A - B) + \cos(A + B))/2$ and filtering out the high-frequency component $\cos(A + B)$, v_m is then simplified as $v_m = \cos\left(2\pi\left(f_c\tau - \frac{B(\tau^2 - 2t\tau)}{2T}\right)\right)$ (the green line in Fig. 5.35). The frequency of v_m can be represented as $f_b = \frac{1}{2\pi} \times \frac{\delta \text{ phase}(v_m)}{\delta t} = \frac{2f_c v}{C} + \frac{2BR}{CT} + \frac{4Bvt}{CT} - \frac{4Bv^2t + 4BRv}{C^2T}$. The terms with respect to $1/C^2$ are too small and can be ignored; thus, Δf can be then simplified as $f_b = \frac{2f_c v}{C} + \frac{2BR}{CT} + \frac{4Bvt}{CT}$. For the static or very-slow-moving target, i.e., v is close to 0, we can further simplify the frequency of mixed signal, and thus obtain the distance between transceiver and target as

$$R = \frac{CT}{2B} f_b. \tag{5.32}$$

Theoretical Range Resolution Upper Bound of FMCW: The transmitted signal v_{tx} has the fundamental frequencies at all multiples of the frequency $1/T$ [208, 209]. If ignoring the amplitude attenuation and assuming that the target is static or moving slowly, the received signal v_{rx} is a time-shifted version of transmitted signal v_{tx}, so v_{tx} and v_{rx} have the same fundamental frequencies. Thus, $v_m = v_{tx} \cdot v_{rx}$ has the fundamental frequencies at all multiples of the frequency $1/T$, since the periodicity is preserved when two same period signals are multiplied [208]. It means that in the frequency domain, v_m has the spectral points only at $1/T$ Hz. In other words, the frequency resolution of mixed signal is $\delta f_b \geq 1/T$. Thereby, we can infer the theoretical range resolution of FMCW as

$$\delta R = \frac{CT}{2B} \delta f_b, \tag{5.33}$$

which indicates that the range resolution, i.e., fixed error, δR depends on frequency resolution of the mixed signal δf_b. However, as mentioned above, δf_b is bounded by the chirp frequency $1/T$, i.e., $\delta f_b \geq 1/T$. In other words, in order to distinguish targets in two different distances, the frequency of the mixed signal cannot be smaller than the chirp frequency [209]. Thus, the theoretical range resolution upper bound of FMCW is computed as

$$\delta R \geq \frac{CT}{2B} \cdot \frac{1}{T} = \frac{C}{2B}, \tag{5.34}$$

which indicates that the range resolution of linear modulated FMCW depends on sweep bandwidth B. Narrower bandwidth results in lower range resolution.

Highest Range Resolution of FMCW Using Commodity Audio Devices: There are two practical frequency bounds that restrict the available sweeping bandwidth B (as shown in Fig. 5.35). On the one hand, the upper bound of human audible range is about 18 kHz. To be inaudible to users, the starting frequency of chirp signal f_c (as shown in Fig. 5.35) should be set higher than 18 kHz. On the other hand, for commodity audio devices, the frequency response starts to decrease rapidly in the band higher than 23 kHz. Thus, the available sweeping bandwidth $B = 23$ kHz $-$ 18 kHz $= 5$ kHz. Accordingly, the highest range resolution of FMCW using commodity audio devices is represented as follows:

$$\delta R \geq \frac{C}{2B} = \frac{343}{2 \times 5000} = 0.0343 \text{ m} = 3.43 \text{ cm}, \qquad (5.35)$$

in which 3.43 cm is much larger than the typical chest displacement of 1 cm during breathing. It implies that directly implementing FMCW using commodity audio devices cannot reliably detect human respiration.

5.6.3.2 Cross-Correlation Function-Based FMCW (C-FMCW)

Basic Idea of C-FMCW: The objective of distance estimation is to measure delay τ. For transmitted signal with a modulation period T, the delay within $T/2$ reflected from target can be measured using the main peak of cross-correlation of transmitted signal and received signal. In this chapter, we use FMCW wave as transmitted signal, and exploit cross-correlation to measure the delay. We call this method cross-correlation function-based FMCW (C-FMCW). Specifically, the cross-correlation function is defined as

$$R(n) = \begin{cases} \dfrac{1}{N-n} \sum_{m=0}^{N-n-1} v_{tx}(m) \cdot v_{rx}(m+n), n \geq 0 \\[4mm] \dfrac{1}{N-|n|} \sum_{m=0}^{N-|n|-1} v_{rx}(m) \cdot v_{tx}(m+n), n < 0 \end{cases}, \qquad (5.36)$$

where N is the number of samples of transmitted signal in one modulation period T, $n = -N+1, -N+2, \ldots, N-1$. $R(n)$ measures the similarity of v_{tx} and n-sample-shifted version of v_{rx}. Suppose $R(n)$ reaches its peak $R(Lag)$ with the Lag-sample shift. Since v_{rx} is a time-shifted version of v_{tx} with amplitude attenuation, it means that v_{rx} overlaps v_{tx} with the Lag-sample shift. Thus, we can measure the time of flight by counting the number of audio samples. For example, Fig. 5.36a–b shows the transmitted signal and received signal with three modulation periods (sampling frequency $F_s = 48$ kHz, $f_c = 18$ kHz, $B = 3$ kHz, $T = 0.08$ s), respectively.

The received signal is set to $Lag = 1047$ sample-shifted version of transmitted signal. Figure 5.36c shows the cross-correlation $R(n)$ of transmitted signal and received signal. We can see the peak in $R(n)$. According to the definition of $R(n)$, the peak should be $R(Lag) = R(1047)$. In Fig. 5.36d, we can see that peak appears indeed at $n = 1047$. Thus, the delay τ can be calculated as

$$\tau = \frac{Lag}{F_s}, \qquad (5.37)$$

where F_s is the sampling frequency. Thereby, we can measure the distance R between transceiver and target as

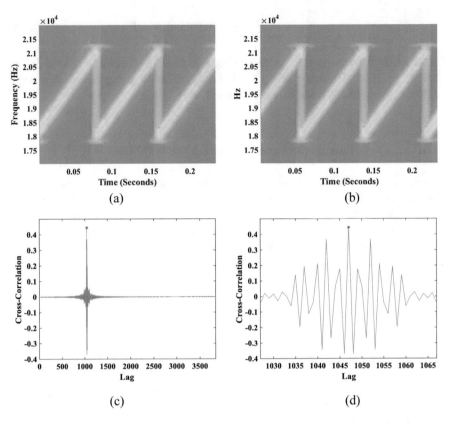

Fig. 5.36 Cross-correlation function of transmitted and received chirp signals with delay. (**a**) Transmitted signal. (**b**) Received signal with delay. (**c**) Cross-correlation function of transmitted and received signals. (**d**) Enlarged view round the peak of (**c**)

$$\begin{cases} \tau = Lag/F_s \\ \tau = 2(R + vt)/C \end{cases} \text{ and } R = \frac{C \cdot Lag}{2F_s} - vt. \qquad (5.38)$$

For the static or very-slow-moving target, i.e., v is close to 0, we can simplify R as

$$R = \frac{C \cdot Lag}{2F_s}. \qquad (5.39)$$

Theoretical Range Resolution Upper Bound of C-FMCW. According to the above equation, the theoretical range resolution of C-FMCW can be represented as

$$\delta R = \frac{C \cdot \delta Lag}{2F_s}. \qquad (5.40)$$

Furthermore, according to the definition of cross-correlation function, the theoretical resolution of *Lag* is $\delta Lag = 1$, since the cross-correlation is calculated at the resolution of each audio sample. Thus, the theoretical range resolution of C-FMCW can be calculated as

$$\delta R = \frac{C \cdot \delta Lag}{2F_s} = \frac{C}{2F_s}. \qquad (5.41)$$

Accordingly, we can see that the range resolution of C-FMCW is limited only by the sampling frequency F_s and independent of sweeping bandwidth B. Higher sampling frequency results in higher ranging resolution. In other words, even if available sweeping bandwidth is narrow, C-FMCW can still achieve very high ranging resolution.

Range Distance Resolution of C-FMCW Using Commodity Audio Devices. Generally, commodity audio systems support 22.05, 44.1, and 48 kHz sampling frequencies. In order to achieve high ranging resolution, we set the sampling frequency to $F_s = 48$ kHz, and the ranging resolution of C-FMCW is calculated as

$$\delta R = \frac{C}{2F_s} = \frac{343}{2 \times 48000} = 0.00357 \text{ m} = 0.357 \text{ cm}. \qquad (5.42)$$

Such a ranging resolution suffices to reliably detect human respiration. In fact, by combining appropriate interpolation techniques, the resolution can be further improved. In this work, we have adopted $2\times$ interpolation method to improve the resolution as follows:

$$\delta R = \frac{C}{2 \times 2F_s} = \frac{343}{2 \times 2 \times 48000} = 0.179 \text{ cm}. \qquad (5.43)$$

5.6.3.3 C-FMCW Verification

In this section, we conduct real experiments to verify C-FMCW. The experimental settings and results are reported as follows.

We bind a general speaker (JBL Jembe, 6 Watt, 80 dB) and microphone (SAMSON MeteorMic, 16 bit, 48 kHz) together (as shown in Fig. 5.37a); they work as a simple acoustic radar (transceiver). The transceiver illuminates target directly (as shown in Fig. 5.37b). The speaker transmits FMCW signal ($F_s = 48$ KHz, $f_c = 18$ KHz, $B = 3$ KHz, $T = 0.02$ s) continuously, which is beyond human audibility range. Meanwhile, the microphone receives the signal reflected by the target at the frequency of 48 kHz with 16 bits.

A ruler is placed on the desk as the ground truth. The position, where the reading of the ruler is 0, is regarded as the reference position (highlighted with blue dashed line in Fig. 5.38b). The distance between transceiver and reference position is set to

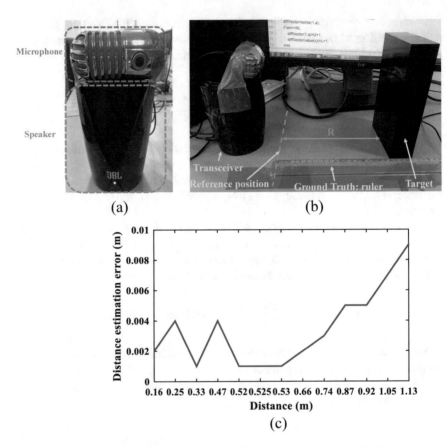

Fig. 5.37 C-FMCW verification. (**a**) Transceiver, (**b**) full view of experiment, (**c**) distance estimation error as ground truth is varied

0. To verify the ranging resolution of C-FMCW, we move the target to 13 different positions ranging from 0.1 to 1.2 m. The distance estimation results are shown in Fig. 5.38c. We observe that within 80 cm, the distance estimated error can be controlled below 0.4 cm. Moreover, the maximum ranging error is less than 0.2 cm when the distance between the acoustic devices and subject is from 40 to 60 cm, meaning that chest displacement which is less than 0.5 cm can be detected using C-FMCW. As distance increases, we observe that error also increases due to degradation of reflected acoustic signal. For more details, readers are encouraged to watch the video at this link: http://dwz.cn/6EE2Dx.

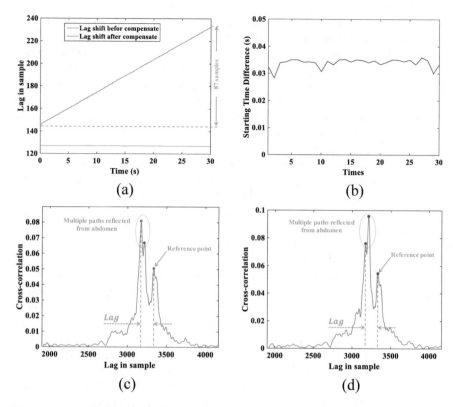

Fig. 5.38 Practical issues using C-FMCW to detect respiration with commodity acoustic devices. (**a**) Lag shift over time caused by sampling frequency offset between transmitter and receiver. The red line indicates the delay shift over time before correction, and the blue line indicates the delay shift over time after correction; (**b**) starting time difference between speaker and microphone; (**c**) multiple reflection path from abdomen; and (**d**) cross-correlation function peak amplitude

5.6.4   Contactless Respiration Detection Using C-FMCW with Commodity Acoustic Devices

In this section, we present the detailed design of C-FMCW-based contactless respiration detection system using commodity acoustic devices. We first identify the technical challenges and our solutions. We then give an overview of the system architecture and describe each major function component in detail.

5.6.4.1   Practical Challenges and Solutions Using C-FMCW to Detect Respiration with Commodity Devices

Even though C-FMCW is promising to achieve high ranging resolution, a number of practical challenges exist when applied for detection of respiration with commodity acoustic devices such as sampling frequency offset between transmitter and receiver, starting time offset between transmitter and receiver, as well as major reflection path selection problems. In this section, we present these challenges and our solutions.

Challenge 1: Sampling Frequency Offset Between Transmitter and Receiver: Existing audio-based ranging approaches typically assume that the sampling frequencies of transmitter and receiver are exactly the same. Unfortunately, due to imperfect clocks, the frequency offset between transmitter and receiver introduces error in calculating *Lag*. Moreover, the error will accumulate over time. In Fig. 5.38a, the red line highlights the *Lag* shift caused by the sampling frequency offset between transmitter and receiver. We find that within 30 s (about 10 breathes), the *Lag* shift accumulates up to 87 samples, which leads to a distance measurement error of 31.1 cm according to the above equations. In other words, during each breathing, the ranging error caused by the sampling frequency offset can exceed 3 cm, which is much larger than the chest displacement of 1 cm during breathing. In order to reliably detect respiration, we must eliminate the errors caused by the sampling frequency offset between transmitter and receiver.

Solution 1: Sampling Frequency Offset Calibration Algorithm. As shown in Fig. 5.39, the basic idea of our solution is to add a shift into transmitted signal to compensate the *Lag* shift caused by sampling frequency offset between transmitter and receiver. Specifically, given the *Lag* shift threshold Td and time duration D, our compensation algorithm will shift audio samples so that the *Lag* shift in duration D is smaller than Td by adapting two parameters, compensation rate r and compensation unit $\Delta Shift$. Compensation rate $r = M/N$ ($M < N$ and both of them are positive integers) means that in every N iterations (as shown in Fig. 5.39), we will add M times shifts into transmitted signal. Compensation unit $\Delta Shift$ means that we add shift in unit $\Delta Shift$ samples. Our compensation algorithm is summarized in Algorithm 5.5.

Algorithm 5.5 Compensation Algorithm

Input: Duration D, and offset threshold OT
Output: Compensation rate r and compensation unit ΔShift

$r \_ L \leftarrow 0; r \_ U \leftarrow 1;$
find the smallest $\Delta Shift$, which makes *getShift*($r \_ L, \Delta Shift, D$) and *getShift* ($r \_ U, \Delta Shift, D$) with different symbol;
// *getShift* (r, ΔShift, D) gets the Lag shift in duration D with the compensation rate r and compensation unit ΔShift
while |*shift*| >*OT* **do** // Binary search method to find appropriate *r*
    shift ← *getShift* ($r \leftarrow (r \_ L + r \_ U$)/2, $\Delta Shift, D$);
    if *shift* < 0 **do** // update the upper bound or lower bound according to the symbol of current system shift
        $r \_ U \leftarrow r;$
    else $r \_ L \leftarrow r;$
End

The algorithm firstly finds shift unit ΔShift to make the symbols of system shift when $r = 0$ and $r = 1$ are different. Then, the algorithm uses the binary search method to find appropriate compensation rate r that makes system shift during period

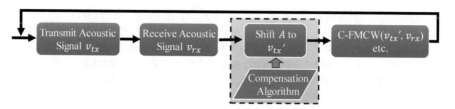

Fig. 5.39 Solution for sampling frequency offset between transmitter and receiver

D smaller than shift threshold OT. Note that during the execution of the Algorithm 5.5, the target should be static. The target movement will introduce shift and influence compensation performance. In practice, we only need to use a static object to get the compensation parameters for once during system calibration before starting respiration detection. The time cost incurred by Algorithm 5.5 depends on the input parameters D and Td. Larger D and smaller Td will result in longer running time but more accurate compensation rate r and compensation unit $\Delta Shift$. The blue line in Fig. 5.38a highlights the system shift after applying our compensation algorithm. We observe that there is almost no *Lag* shift after applying the compensation. In summary, with this compensation algorithm, the effect caused by sampling frequency offset between transmitter and receiver can be well controlled.

Challenge 2: Starting Time Offset Between Transmitter and Receiver: Ideally, if speaker and microphone start to work simultaneously, we can directly estimate the absolute distance from transceiver to abdomen. Unfortunately, even though we send the starting commands to speaker and microphone simultaneously, the starting time of speaker and microphone can be different in practice. Moreover, the starting time difference varies over time. Figure 5.38b shows 30 times starting time difference between speaker and microphone in our system. The varied starting time difference will introduce a large error in initial distance estimation. In the example shown in Fig. 5.38b the starting time difference (average 0.034 s) will introduce $0.034 \times 48000 = 1632$ samples *Lag* shift. According to C-FMCW distance estimation equation, it will introduce an absolute distance error of $343 \times 1632/(2 \times 48000) = 5.83\ m$.

Solution 2: Cancelling Starting Time Offset Using Self-Interference Effect. As shown in Fig. 5.37a, in our system, speaker and microphone are co-located. As a result, part of the transmitted signal will be received by the microphone without reflection (i.e., self-interference). This part of received signal will generate a special peak in the cross-correlation function of transmitted signal and received signal (the reference point shown in Fig. 5.38c,d). Due to no reflection, this part of received signal will not be affected by the target. Thus, both the amplitude and location of this special peak will be kept relatively static. It is easy to use variation of amplitude and location to distinguish this special peak from others. We name the special peak *reference point*. As shown in Fig. 5.38c, d, with the reference point, we can cancel the starting time offset between transmitter and receiver by using the location difference between reference point and the peak generated by the receive signal reflected from the target. Then, by using C-FMCW distance estimation equation, we can accurately get the distance. This solution can completely avoid the interference from the starting time difference between speaker and microphone, since the *Lag* shift caused by the starting time difference is the same for the reference point and the peak generated by the receive signal reflected from the target. Thus, when we calculate the *Lag* by using the location difference between the reference point and the peak generated by the receive signal reflected from abdomen, the *Lag* shift caused by the starting time difference can be cancelled.

Challenge 3: Multiple Reflection Paths from Different Parts of Body: In practice, there are multiple reflection paths from different parts of abdomen. As shown in both Fig. 5.38c and Fig. 5.38d, there are two peaks (in the blue circle) corresponding to two different reflection paths (note that in practice, there may be more than two peaks). Different parts of abdomen have different movement displacements during breathing. The reflection paths from static parts cannot be used for respiration detection. Only the reflection paths from the moving parts during breathing can contribute to respiration detection. Moreover, the power of received signal from different reflection path fluctuates dynamically, which will result in amplitude fluctuation of cross-correlation peaks. As shown in Fig. 5.38c, d, the two peaks in blue circle alternatively become the largest peak. It is challenging to identify the representative peak that can effectively reflect human respiration among the multiple peaks.

Solution 3: Selecting Major Reflection Path Using the Periodicity of Respiration. Even though there may exist multiple reflection paths from different parts of abdomen, we know that human respiration is almost periodical in nature. It means that the distances from the moving parts of the target during breathing will vary periodically, while other distances from static reflection paths are not periodical. Thus, we can select the effective reflection path by measuring the periodicity. The periodicity can be easily measured by using an autocorrelation function. For the sequence $X = x_1 x_2 \ldots x_M$, autocorrelation function is defined as follows:

$$R(k) = \frac{E[(x_i - \mu)(x_{i+k} - \mu)]}{\sigma^2}, k = 1, 2, \ldots, M - 1. \tag{5.44}$$

Figure 5.40a shows two different distance estimations from two different reflection paths over time. Figure 5.40b shows the autocorrelation functions of the two distance estimations, based on which we see that the peaks of the autocorrelation function are higher if distance estimation is more cyclical.

5.6.4.2 C-FMCW-Based Respiration Detection System Framework

In this section, we present the key workflow of C-FMCW-based contactless respiration detection system using acoustic signal. We first give an overview of the system architecture, and then describe each key functional module in detail.

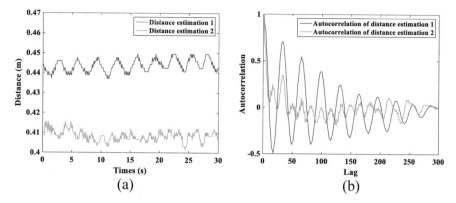

Fig. 5.40 Autocorrelations of two different distance estimation results. (**a**) Two distance estimation sequences. (**b**) The autocorrelation function of the two distance estimation sequences in (**a**)

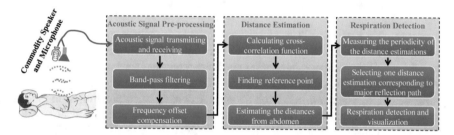

Fig. 5.41 Overall framework of C-FMCW-based contactless respiration detection system using acoustic signal

System Architecture: As shown in Fig. 5.41, our system is composed of three major components, namely acoustic signal preprocessing, distance estimation, and respiration detection. The acoustic signal preprocessing module filters out noise and compensates the *Lag* shift. The distance estimation module estimates the distance of each reflection path from the target. The respiration detection module detects the respiration from the distance estimation results. In the following, we summarize the key functional modules of C-FMCW.

Acoustic Signal Preprocessing: The objectives of this module are to (1) ensure real-time continuous transmission and reception of acoustic signals; (2) denoise received signals; and (3) compensate the *Lag* shift caused by sampling frequency offset between transmitter and receiver.

In order to filter background noise, we apply a band-pass filter (passband $[f_c, f_c + B]$), where f_c denotes carrier frequency and B denotes sweep bandwidth.

To compensate the *Lag* shift caused by sampling frequency offset between transmitter and receiver, we firstly get the compensation rate r and compensation unit $\Delta Shift$ by using our composed compensation algorithm in Algorithm 5.5. Then, we distribute M times compensation (each time ΔShift samples) to transmitted signal (as shown in Fig. 5.39) during N iterations uniformly to counteract the shift caused by the sampling frequency offset between transmitter and receiver. M and N have the relation $M/N = r$.

Distance Estimation: The key function of this module is to estimate the distance of each reflection path from abdomen using our proposed C-FMCW. Specifically, we first calculate the cross-correlation function of compensated transmitted signal and received signal, and find all the peaks of cross-correlation function. Then, we identify the reference point by finding the peak whose variations of the amplitude and location are smallest. After the reference point is located, we estimate the absolute distances (here the distance means half of the length of reflection path) of all reflection paths by using the solution presented in the above section (i.e., acoustic signal preprocessing). We calculate *Lag* by using the location difference between reference points and the peak corresponding to current reflection path. Then, we estimate the distance by using C-FMCW distance estimation equation. The distance estimation result of each reflection path over time is saved as a distance sequence. These distance sequences of all the reflection paths will be regarded as the input of next module.

Respiration Detection: The major work of this module is to detect respiration based on the distance estimation results. As mentioned in Challenge 2 of the above section not all the distance estimation results can be used for respiration detection. We first calculate the periodicity of all the distance sequences by using autocorrelation function. Then, we select the distance sequence corresponding to the major reflection path which can reflect the respiration well if the periodicities of all the distance sequences are lower than a given threshold. Thereby, the interference factors (e.g., body movement, sleep posture change) can be detected. Figure 5.42 shows the respiration detection result while the user changes sleep posture. The system will stop identifying breathing when the user changes sleep posture, since during this period the distance estimation fluctuation cannot reliably capture the chest movement. The system will recover to identify breathing once at least one distance sequence's periodicity becomes higher than a threshold.

5.6.5 Performance Evaluation

To evaluate the proposed system, we used two laptops (Thinkpad T450 with Intel Core i5–5200 CPU, 8G RAM; Dell Latitude E6540 with Intel Core i7-4800MQ, 4GB RAM) and connected them with a pair of commodity speaker (JBL Jembe, 6 Watt, 80 dB) and microphone (SAMSON MeteorMic, 16 bit, 48 KHz), which forms an audio transceiver in our system, as shown in Fig. 5.37a. In particular, we used two transceivers to increase the coverage of the target monitoring area.

We used two representative audio-based approaches [201, 202] as baseline methods, and performed comprehensive experiments to evaluate the system in four different rooms with 22 subjects, who have three different sleep postures. Specifically, while we also have evaluated the system's robustness with different

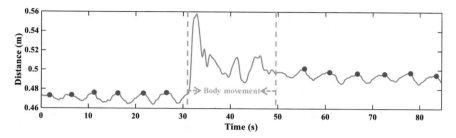

Fig. 5.42 Respiration identification with body movement

scenarios, different sensing distances, different respiration rates, different kinds of simulated apneas, and body movements, we will not present the results in this section for the sake of simplicity. For more details about the experimental setup and results, please refer to the article [210].

5.6.5.1   Evaluation with Different Subjects

The respiration detection performance is shown in Fig. 5.43a. We can see that the median respiration detection error of the proposed system is lower than 0.35 breaths/ min, while the median error of the baseline approach is around 1 breaths/min. In addition, the maximum error of the proposed system is about 0.6 breaths/min, while the corresponding value of the baseline reaches 3.7 breaths/min. This experimental result indicates that (1) the proposed C-FMCW method achieves a very high distance estimation resolution and (2) the system can accurately detect human respiration when the participants sleep on his/her back.

5.6.5.2   Evaluation with Different Sleep Postures

Except for sleeping on one's back, lying on one side is also a common sleep posture. With the same experiment settings, all the subjects are recruited to evaluate the performance of the system when lying on one side. Figure 5.43b, c shows the respiration detection error CDF when subjects are sleeping on the left and right sides, respectively. We can find that the median error of the proposed system is 0 breaths/min, while the median error of the baseline is 0.35 breaths/min. In addition, the maximum error of our system can be controlled under 0.35 breaths/min, while the corresponding value of the baseline is 1 breaths/min. To sum up, the proposed system is able to accurately detect human respiration for all three common sleep postures and achieve better performance.

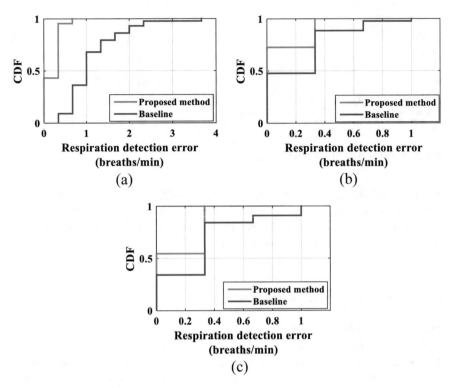

Fig. 5.43 Respiration detection results of 22 subjects with three different sleep postures. (**a**) Sleep on one's back, (**b**) facing left, (**c**) facing right

References

1. S. D. Nagowah, "Aiding social interaction via a mobile peer to peer network," in Proc. 4th Int. Conf. Digital Soc., 2010, pp. 130–135.
2. T. Liu, B. Paramvir, and C. Imrich, "Mobility modeling, location tracking, and trajectory prediction in wireless ATM networks," IEEE J. Sel. Areas Commun., vol. 16, no. 6, pp. 922–936, Aug. 1998.
3. J. Lawrence, T. R. Payne, and R. V. Kripalani, "Exploiting incidental interactions between mobile devices," MIRW, vol. 56, pp. 56–59, 2006.
4. I. Buchem, "Serendipitous learning: recognizing and fostering the potential of microblogging," Formare Open J. per la formazione rete, vol. 11, no. 74, pp. 7–16, 2011.
5. C. Brown, C. Efstratiou, I. Leontiadis, D. Quercia, and C. Mascolo, "Tracking serendipitous interactions: How individual cultures shape the office," in Proc. 17th ACM Conf. Comput. Supported Cooperative Work Social Comput., 2014, pp. 1072–1081.
6. C. M. Huang, K. C. Lan, and C. Z. Tsai, "A survey of opportunistic networks," in Proc. 22nd Int. Conf. Adv. Inf. Netw. Appl., 2008, pp. 1672–1677.

7. G. Yavas̨, D. Katsaros, O. Ulusoy, and Y. Manolopoulos, "A data mining approach for location prediction in mobile environments," Data Knowl. Eng., vol. 54, no. 2, pp. 121–146, 2005.

8. B. Thoms, "A dynamic social feedback system to support learning and social interaction in higher education," IEEE Trans. Learning Technol., vol. 4, no. 4, pp. 340–352, Oct.–Dec. 2011.

9. R. Beale, "Supporting social interaction with smart phones," IEEE Pervasive Comput., vol. 4, no. 2, pp. 35–41, Jan.–Mar. 2005.

10. W. S. Yang, S. Y. Hwang, and Y. W. Shih, "Facilitating information sharing and social interaction in mobile peer-to-peer environment," in Proc. Pervasive Comput. Commun. Workshops, 2012, pp. 673–678.

11. P. Tamarit, C. T. Calafate, J. C. Cano, and P. Manzoni, "BlueFriend: Using Bluetooth technology for mobile social networking," in Proc. 6th Annu. Int. Conf. Mobile Ubiquitous Syst.: Comput., Netw. Services, 2009, pp. 1–2.

12. J. A. Paradiso, J. Gips, M. Laibowitz, S. Sadi, D. Merrill, R. Aylward, P. Maes, and A. Pentland, "Identifying and facilitating social interaction with a wearable wireless sensor network," Pers. Ubiquitous Comput., vol. 14, no. 2, pp. 137–152, 2010.

13. H. Xiong, D. Zhang, D. Zhang, and V. Gauthier, "Predicting mobile phone user locations by exploiting collective behavioral patterns," in Proc. 9th Int. Conf. Ubiquitous Intell. Comput., 2012, pp. 164–171.

14. T. M. T. Do and D. Gatica-Perez, "Contextual conditional models for smartphone-based human mobility prediction" in Proc. ACM Conf. Ubiquitous Comput., 2012, pp. 163–172.

15. E. Cho, S. A. Myers, and J. Leskovec, "Friendship and mobility: User movement in location-based social networks," in Proc. 17th ACM SIGKDD Int. Conf. Knowl. Discovery Data Mining, 2011, pp. 1082–1090.

16. A. Noulas, S. Scellato, N. Lathia, and C. Mascolo, "Mining user mobility features for next place prediction in location-based services," in Proc. IEEE 12th Int. Conf. Data Mining, 2012, pp. 1038–1043.

17. P. Baumann, W. Kleiminger, and S. Santini, "The influence of temporal and spatial features on the performance of next-place prediction algorithms," in Proc. ACM Int. Joint Conf. Pervasive Ubiquitous Comput., 2013, pp. 449–458.

18. J. McInerney, J. Zheng, A. Rogers, and N. R. Jennings, "Modelling heterogeneous location habits in human populations for location prediction under data sparsity," in Proc. ACM Int. Joint Conf. Pervasive Ubiquitous Comput., 2013, pp. 469–478.

19. C. Song, Z. Qu, N. Blumm, and A.-L. Barabsi, "Limits of predictability in human mobility," Science, vol. 327, no. 5968, pp. 1018–1021, 2010.

20. M. Lin, W. J. Hsu, and Z. Q. Lee, "Predictability of individuals' mobility with high-resolution positioning data," in Proc. ACM Int. Joint Conf. Pervasive Ubiquitous Comput., 2012, pp. 381–390.

21. G. Chen, S. Hoteit, A. C. Viana, M. Fiore, and C. Sarraute, "Enriching sparse mobility information in Call Detail Records," Computer Communications, vol. 122, pp. 44–58, 2018.

22. L. Wang, Z. Yu, B. Guo, T. Ku, and F. Yi, "Moving destination prediction using sparse dataset: A mobility gradient descent approach," ACM Transactions on Knowledge Discovery from Data (TKDD), vol. 11, no. 3, pp. 37, 2017.

23. X. Song, R. Shibasaki, N. J. Yuan, X. Xie, T. Li, and R. Adachi, "Deep Mob: learning deep knowledge of human emergency behavior and mobility from big and heterogeneous data," ACM Transactions on Information Systems (TOIS), vol. 35, no. 4, pp. 41, 2017.

24. R. Jiang, X. Song, Z. Fan, T. Xia, Q. Chen, Q. Chen, and R. Shibasaki, "Deep ROI-Based Modeling for Urban Human Mobility Prediction," Proceedings of the ACM on Interactive, Mobile, Wearable and Ubiquitous Technologies, vol. 2, no. 1, pp. 14, 2018.

25. B. Resch, F. Calabrese, A. Biderman, and C. Ratti, "An approach towards real-time data exchange platform system architecture," in Proc. 6th Annu. IEEE Int. Conf. Pervasive Comput. Commun., 2008, pp. 153–159.
26. E. Toch and I. Levi, "Locality and privacy in people-nearby applications," in Proc. ACM Int. Joint Conf. Pervasive Ubiquitous Comput., 2013, pp. 539–548.
27. J. Quinlan, C4.5: Programs for Machine Learning. San Mateo, CA, USA: Morgan Kaufmann, 1993.
28. J. Quinlan, "Introduction of decision tree," Mach. Learning, vol. 1, no. 1, pp. 81–106, 1986.
29. I. Witten, Data Mining—Practical Machine Learning Tools and Techniques, 2nd ed., San Francisco, CA, USA: Morgan Kaufmann, 2005.
30. N. Friedman, D. Geiger, and M. Goldszmidt, "Bayesian network classifiers," Mach. Learning, vol. 29, pp. 131–163, 1997.
31. M. Hall, E. Frank, G. Holmes, B. Pfahringer, P. Reutemann, and I. Witten, "The WEKA data mining software: An update," ACM SIGKDD Explorations Newslett., vol. 11, no. 1, pp. 10–18, 2009.
32. J. Quinlan, "Learning with continuous classes," in Proc. Australian Joint Conf. Artif. Intell., 1992, pp. 343–348.
33. L. Nguyen, H. T. Cheng, P. Wu, S. Buthpitiya, and Y. Zhang, "PnLUM: System for prediction of next location for users with mobility," in Proc. Nokia Mobile Data Challenge Workshop, vol. 2, 2012.
34. Z. Yu, H. Wang, B. Guo, T. Gu and T. Mei, "Supporting Serendipitous Social Interaction Using Human Mobility Prediction," in IEEE Transactions on Human-Machine Systems, vol. 45, 6, 811–818, Dec. 2015. DOI: https://doi.org/10.1109/THMS.2015.2451515
35. UNFPA, State of World Population 2011. [Online] Available: http://foweb.unfpa.org/SWP2011/reports/EN-SWOP2011-FINAL.pdf.
36. C. Kawas, R. Katzman, Epidemiology of dementia and Alzheimer's disease, Alzheimer Dis. (1999) 95–116.
37. K. Du, D. Zhang, X. Zhou, M. Hariz, Handling conflicts of context-aware reminding system in sensorised home, Cluster Comput. 14 (1) (2011) 81–89.
38. T. Giovannetti, D.J. Libon, L.J. Buxbaum, M.F. Schwartz, Naturalistic action impairments in dementia, Neuropsychologia 40 (8) (2002) 1220–1232.
39. B. Reisberg, S. Ferris, M. Leon, T. Crook, The global deterioration scale for assessment of primary degenerative dementia, Am. J. Psychiatry 139 (1982) 1136–1139.
40. R. McShane, K. Gedling, J. Keene, et al., Getting lost in dementia: a longitudinal study of a behavioural symptom, Int. Psychogeriatr. 10 (1998) 253–260.
41. nedap-securitymanagement [Online] Available: http://www.nedap-securitymanagement.com/.
42. L. Robinson, D. Hutchings, L. Corner, F. Beyer, et al., A systematic literature review of the effectiveness of non-pharmacological interventions to prevent wandering in dementia and evaluation of the ethical implications and acceptability of their use, Health Technol. Assess. 10 (26) (2006) 1–124.
43. H. Ogawa, Y. Yonezawa, H. Maki, et al. A mobile phone-based safety support system for wandering elderly persons, in: Proc. of Engineering in Medicine and Biology Society, EMBS, 2004, pp. 3316–3317.
44. D. Patterson, L. Liao, K. Gajos, et al., Opportunity knocks: a system to provide cognitive assistance with transportation services, in: Proc. of UbiComp, 2004, in: LNCS, vol. 3205, 2004, pp. 433–450.
45. Y. Chang, Anomaly detection for travelling individuals with cognitive impairments, Newsl. ACM SIGACCESS Access. Comput. 97 (2010) 25–32.
46. M.C. González, C.A. Hidalgo, A.L. Barabási, Understanding individual human mobility patterns, Nature 453 (2008) 779–782.
47. F.T. Liu, K.M. Ting, Z.-H. Zhou, Isolation forest, in: 8th IEEE International Conference on Data Mining, ICDM, pp. 413–422.

48. K. Shimizu, K. Kawamura, K. Yamamoto, Location system for dementia wandering, in: Proc. of EMBS, 2000, pp. 1556–1559.
49. S. Matsuoka, H. Ogawa, H. Maki, et al. A new safety support system for wandering elderly persons, in: Proc. of EMBS, 2011, pp. 5232–5235.
50. M. Mulvenna, S. Sävenstedt, F. Meiland, et al. Designing & evaluating a cognitive prosthetic for people with mild dementia, in: Proc. of ECCE, 2010, pp. 11–18.
51. C. Palomino, P. Heras-Quiros, et al. Outdoors monitoring of elderly people assisted by compass, GPS and mobile social network, in: Proc. of IWANN, 2009, pp. 808–811.
52. J. Brush, M. Calkins, Cognitive impairment, wayfinding, and the long-term care environment, Pers. Gerontol. 13 (2) (2008) 65–73.
53. L. Liao, D. Fox, H. Kautz, Learning and inferring transportation routines, in: Proc. of AAAI, 2004, pp. 348–353.
54. Y. Chang, T. Wang, Mobile location-based social networking in supported employment for people with cognitive impairments, Cybern. Syst. 41 (3) (2010) 245–261.
55. J.-G. Lee, J. Han, X. Li, Trajectory outlier detection: a partition-and-detect framework, in: Proc. of ICDE, 2008, pp. 140–149.
56. Y. Ge, H. Xiong, Z.-H. Zhou, et al. Top-eye: top-k evolving trajectory outlier detection, in: Proc. of CIKM, 2010, pp. 1733–1736.
57. Y. Bu, L. Chen, A.W.-C. Fu, D. Liu, Efficient anomaly monitoring over moving object trajectory streams, in: Proc. of KDD, 2009, pp. 159–168.
58. X. Li, Z. Li, J. Han, J.-G. Lee, Temporal outlier detection in vehicle traffic data, in: Proc. of ICDE, 2009, pp. 1319–1322.
59. X. Li, J. Han, S. Kim, H. Gonzalez, ROAM: rule- and motif-based anomaly detection in massive moving object data sets, in: Proc. of SDM, 2007, pp. 273–284.
60. R.R. Sillito, R.B. Fisher, Semi-supervised learning for anomalous trajectory detection, in: Proc. of BMVC, 2008, pp. 1035–1044.
61. Z. Liao, Y. Yu, B. Chen, Anomaly detection in GPS data based on visual analytics, in: Proc. of VAST, 2010, pp. 51–58.
62. D. Zhang, N. Li, Z. Zhou, C. Chen, et al. iBAT: detecting anomalous taxi trajectories from GPS traces, in: Proc. of UbiComp, 2011, pp. 99–108.
63. C. Chen, D. Zhang, P.S. Castro, et al. Real-time detection of anomalous taxi trajectories from GPS traces, in: Proc. of MobiQuitous, 2011, pp. 1–12.
64. J. Yuan, Y. Zheng, X. Xie, G. Sun, T-drive: enhancing driving directions with taxi drivers' intelligence, IEEE Trans. Knowl. Data Eng. 25 (1) (2013)220–232.
65. C. Parent, S. Spaccapietra, C. Renso, et al., Semantic trajectories modeling and analysis, ACM Comput. Surv. 45 (4) (2013) 42–64.
66. D. Martino-Saltzman, B.B. Blasch, R.D. Morris, et al., Travel behavior of nursing home residents perceived as wanderers and nonwanderers, Gerontologist 31 (5) (1991) 666–672.
67. Y. Zheng, Q. Li, Y. Chen, et al. Understanding mobility based on GPS data, in: Proc. of UbiComp, 2008, pp. 312–321.
68. Y. Zheng, X. Xie, W. Ma, GeoLife: a collaborative social networking service among user, location and trajectory, IEEE Data Eng. Bull. 33 (2) (2010) 32–40.
69. Q. Lin, D. Zhang, K. Connelly, H. Ni, Z. Yu, X. Zhou, Disorientation detection by mining GPS trajectories for cognitively-impaired elders, Pervasive and Mobile Computing, 19, 2015, 71–85
70. K. Marshall, "Working with computers," Perspectives Labour Income, vol. 22, no. 5, pp. 9–15, 2001.
71. E. A. Boyle, T. M. Connolly, T. Hainey, and J. M. Boyle, "Engagement in digital entertainment games: A systematic review," Comput. Human Behav., vol. 28, no. 3, pp. 771–780, 2012.
72. B. Schilit, N. Adams, and R. Want, "Context-aware computing applications," in Proc. 1st Workshop Mobile Comput. Syst. Appl. (WMCSA), Santa Cruz, CA, USA, 1994, pp. 85–90.

73. M. Griffiths, "Does Internet and computer 'addiction' exist? Some case study evidence," CyberPsychol. Behav., vol. 3, no. 2, pp. 211–218, 2000.

74. L. Chen, J. Hoey, C. D. Nugent, D. J. Cook, and Z. Yu, "Sensor-based activity recognition," IEEE Trans. Syst., Man, Cybern. C, Appl. Rev., vol. 42, no. 6, pp. 790–808, Nov. 2012.

75. B. Guo, Z. Yu, L. Chen, X. Zhou, and X. Ma, "Mobigroup: Enabling lifecycle support to social activity organization and suggestion with mobile crowd sensing," IEEE Trans. Human–Mach. Syst., vol. 46, no. 3, pp. 390–402, Jun. 2016.

76. R. Piyare, "Internet of Things: Ubiquitous home control and monitoring system using android based smart phone," Int. J. Internet Things, vol. 2, no. 1, pp. 5–11, 2013.

77. A. Mehrotra, R. Hendley, and M. Musolesi, "Prefminer: Mining user's preferences for intelligent mobile notification management," in Proc. UBICOMP, Heidelberg, Germany, 2016, pp. 1223–1234.

78. Z. Yu, H. Xu, Z. Yang, and B. Guo, "Personalized travel package with multi-point-of-interest recommendation based on crowdsourced user footprints," IEEE Trans. Human–Mach Syst., vol. 46, no. 1, pp. 151–158, Feb. 2016.

79. Z. Yu, F. Yi, Q. Lv, and B. Guo, "Identifying on-site users for social events: Mobility, content, and social relationship," IEEE Trans. Mobile Comput., to be published, doi: https://doi.org/10.1109/TMC.2018.2794981.

80. P. Hevesi, S. Wille, G. Pirkl, N. Wehn, and P. Lukowicz, "Monitoring household activities and user location with a cheap, unobtrusive thermal sensor array," in Proc. UBICOMP, 2014, pp. 141–145.

81. E. Thomaz, I. Essa, and G. D. Abowd, "A practical approach for recognizing eating moments with wrist-mounted inertial sensing," in Proc. UBICOMP, Osaka, Japan, 2015, pp. 1029–1040.

82. J. Korpela, R. Miyaji, T. Maekawa, K. Nozaki, and H. Tamagawa, "Evaluating tooth brushing performance with smartphone sound data," in Proc. UBICOMP, Osaka, Japan, 2015, pp. 109–120.

83. T. Hao, G. Xing, and G. Zhou, "RunBuddy: A smartphone system for running rhythm monitoring," in Proc. UBICOMP, Osaka, Japan, 2015, pp. 133–144.

84. P. Marquardt, A. Verma, H. Carter, and P. Traynor, "(sp) iPhone: Decoding vibrations from nearby keyboards using mobile phone accelerometers," in Proc. 18th ACM Conf. Comput. Commun. Security, Chicago, IL, USA, 2011, pp. 551–562.

85. D. Asonov and R. Agrawal, "Keyboard acoustic emanations," in Proc. IEEE Symp. Security Privacy, vol. 2004. Berkeley, CA, USA, 2004, pp. 3–11.

86. J. Wang, K. Zhao, X. Zhang, and C. Peng, "Ubiquitous keyboard for small mobile devices: Harnessing multipath fading for fine-grained keystroke localization," in Proc. MobiSys, Bretton Woods, NH, USA, 2014, pp. 14–27.

87. H. Du et al., "Sensing keyboard input for computer activity recognition with a smartphone," in Proc. ACM Int. Joint Conf. Pervasive Ubiquitous Comput. ACM Int. Symp. Wearable Comput. UbiComp/ISWC, Maui, HI, USA, Sep. 2017, pp. 25–28.

88. H. Jimison, M. Pavel, J. McKanna, and J. Pavel, "Unobtrusive monitoring of computer interactions to detect cognitive status in elders," IEEE Trans. Inf. Technol. Biomed., vol. 8, no. 3, pp. 248–252, Sep. 2004.

89. J. Srivastava, R. Cooley, M. Deshpande, and P.-N. Tan, "Web usage mining: Discovery and applications of usage patterns from Web data," ACM SIGKDD Explor. Newslett., vol. 1, no. 2, pp. 12–23, 2000.

90. H. Du et al., "Group mobility classification and structure recognition using mobile devices," in Proc. PerCom, Sydney, NSW, Australia, 2016, pp. 1–9.

91. B. Guo et al., "Worker-contributed data utility measurement for visual crowdsensing systems," IEEE Trans. Mobile Comput., vol. 16, no. 8, pp. 2379–2391, Aug. 2017.

92. H. Chen, B. Guo, Z. Yu, L. Chen, and X. Ma, "A generic framework for constraint-driven data selection in mobile crowd photographing," IEEE Internet Things J., vol. 4, no. 1, pp. 284–296, Feb. 2017.

93. X. Zhang, W. Li, X. Chen, and S. Lu, "MoodExplorer: Towards Compound Emotion Detection via Smartphone Sensing," Proceedings of the ACM on Interactive, Mobile, Wearable and Ubiquitous Technologies, vol. 1, no. 4, pp. 176, 2018.

94. P. Siirtola and J. Röning, "Recognizing human activities user independently on smartphones based on accelerometer data," Int. J. Interact. Multimedia Artif. Intell., vol. 1, no. 5, pp. 38–45, 2012.

95. A. M. Khan, A. Tufail, A. M. Khattak, and T. H. Laine, "Activity recognition on smartphones via sensor-fusion and KDA-based SVMS," Int. J. Distrib. Sensor Netw., vol. 10, no. 5, pp. 1–14, 2014.

96. M. Shoaib, S. Bosch, Ö. D. Incel, H. Scholten, and P. J. M. Havinga, "Fusion of smartphone motion sensors for physical activity recognition," Sensors, vol. 14, no. 6, pp. 10146–10176, 2014.

97. R. Mohamed and M. Youssef, "Heartsense: Ubiquitous accurate multi-modal fusion-based heart rate estimation using smartphones," Proceedings of the ACM on Interactive, Mobile Wearable and Ubiquitous Technologies, vol. 1, no. 3, pp. 97, 2017.

98. H. Lu, W. Pan, N. D. Lane, T. Choudhury, and A. T. Campbell, "Soundsense: Scalable sound sensing for people-centric applications on mobile phones," in Proc. MobiSys, 2009, pp. 165–178.

99. H. Du et al., "Recognition of group mobility level and group structure with mobile devices," IEEE Trans. Mobile Comput., to be published, doi: https://doi.org/10.1109/TMC.2017.2694839.

100. X. Sun, Z. Lu, W. Hu, and G. Cao, "Symdetector: Detecting sound-related respiratory symptoms using smartphones," in Proc. UbiComp, Osaka, Japan, 2015, pp. 97–108.

101. T. Zhu, Q. Ma, S. Zhang, and Y. Liu, "Context-free attacks using keyboard acoustic emanations," in Proc. ACM SIGSAC Conf. Comput. Commun. Security, Scottsdale, AZ, USA, 2014, pp. 453–464.

102. J. Liu et al., "Snooping keystrokes with mm-level audio ranging on a single phone," in Proc. MobiCom, Paris, France, 2015, pp. 142–154.

103. B. Chen, V. Yenamandra, and K. Srinivasan, "Tracking keystrokes using wireless signals," in Proc. MobiSys, Florence, Italy, 2015, pp. 31–44.

104. K. Ali, A. X. Liu, W. Wang, and M. Shahzad, "Recognizing keystrokes using WiFi devices," IEEE Journal on Selected Areas in Communications, vol. 35, no. 5, pp. 1175–1190, 2017.

105. L. Zhuang, F. Zhou, and J. D. Tygar, "Keyboard acoustic emanations revisited," ACM Trans. Inf. Syst. Security (TISSEC), vol. 13, no. 1, 2009, Art. no. 3.

106. Y. Berger, A. Wool, and A. Yeredor, "Dictionary attacks using keyboard acoustic emanations," in Proc. 13th ACM Conf. Comput. Commun. Security, Alexandria, VA, USA, 2006, pp. 245–254.

107. W. K. Pratt, "Generalized Wiener filtering computation techniques," IEEE Trans. Comput., vol. C-21, no. 7, pp. 636–641, Jul. 1972.

108. R.-Z. Zhang and H.-J. Cui, "Speech endpoint detection algorithm analyses based on short-term energy," Audio Eng., vol. 7, no. 7, p. 015, 2005.

109. M. Damashek, "Gauging similarity with n-Grams: Language-independent categorization of text," Science, vol. 267, no. 5199, pp. 843–848, 1995.

110. J. Kaur and J. R. Saini, "Emotion detection and sentiment analysis in text corpus: A differential study with informal and formal writing styles," Int. J. Comput. Appl., vol. 101, no. 9, pp. 1–9, 2014.

111. P. H. Dietz, B. Eidelson, J. Westhues, and S. Bathiche, "A practical pressure sensitive computer keyboard," in Proc. 22nd Annu. ACM Symp. User Interface Softw. Technol., Victoria, BC, Canada, 2009, pp. 55–58.

112. W. van den Hoogen, E. Braad, and W. A. IJsselsteijn, "Pressure at play: Measuring player approach and avoidance behaviour through the keyboard," in Proc. DIGRA Int. Conf., vol. 2014, no. 8. Salt Lake City, UT, USA, 2014, p. 12.

113. Z. Yu, H. Du, D. Xiao, Z. Wang, Q. Han and B. Guo, "Recognition of Human Computer Operations Based on Keystroke Sensing by Smartphone Microphone," in IEEE Internet of Things Journal, vol. 5, 2, pp. 1156–1168, April 2018.

114. World Health Organization. Drowning fact sheet number 347. 2010.

115. Eng, How-Lung, et al. "DEWS: a live visual surveillance system for early drowning detection at pool." Circuits and Systems for Video Technology, IEEE Transactions on 18.2 (2008): 196–210.

116. Kharrat, Mohamed, et al. "Near drowning pattern recognition using neural network and wearable pressure and inertial sensors attached at swimmer's chest level." Mechatronics and Machine Vision in Practice (M2VIP), 2012 19th International Conference. IEEE, 2012.

117. iSwimband Wearable Drowning Detection Device. https://www.iswimband.com/

118. Martin E, Vinyals O, Friedland G, et al. Precise indoor localization using smart phones. Proceedings of the international conference on Multimedia. ACM, 2010: 787–790.

119. Xiong J, Jamieson K. ArrayTrack: A Fine-Grained Indoor Location System. NSDI. 2013: 71–84.

120. Hsu H H, Peng W J, Shih T K, et al. Smartphone Indoor Localization with Accelerometer and Gyroscope. Network-Based Information Systems (NBiS), 2014 17th International Conference on. IEEE, 2014: 465–469.

121. Qian J, Ma J, Ying R, et al. An improved indoor localization method using smartphone inertial sensors. Indoor Positioning and Indoor Navigation (IPIN), 2013 International Conference on. IEEE, 2013: 1–7.

122. Haverinen J, Kemppainen A. Global indoor self-localization based on the ambient magnetic field. Robotics and Autonomous Systems, 2009, 57(10): 1028–1035.

123. Hassan, M., Daiber, F., Wiehr, F., Kosmalla, F., & Krüger, A. Footstriker: An EMS-based foot strike assistant for running. Proceedings of the ACM on Interactive, Mobile, Wearable and Ubiquitous Technologies, 1(1), 2, 2017.

124. Chun, K. S., Bhattacharya, S., and Thomaz, E. Detecting Eating Episodes by Tracking Jawbone Movements with a Non-Contact Wearable Sensor. Proceedings of the ACM on Interactive, Mobile, Wearable and Ubiquitous Technologies, 2(1), 4, 2018.

125. Seuter, M., Pfeiffer, M., Bauer, G., Zentgraf, K., & Kray, C. Running with Technology: Evaluating the Impact of Interacting with Wearable Devices on Running Movement. Proceedings of the ACM on Interactive, Mobile, Wearable and Ubiquitous Technologies, 1(3), 101, 2017.

126. Auvinet B, Gloria E, Renault G, et al. Runner's stride analysis: comparison of kinematic and kinetic analyses under field conditions. Science & Sports, 2002, 17(2): 92–94.

127. Fitzpatrick K, Anderson R. Validation of accelerometers and gyroscopes to provide real-time kinematic data for golf analysis. The Engineering of Sport 6. Springer New York, 2006: 155–160.

128. Spelmezan D, Borchers J. Real-time snowboard training system. CHI'08 Extended Abstracts on Human Factors in Computing Systems. ACM, 2008: 3327–3332.

129. Echterhoff, J. M., Haladjian, J., & Brügge, B. Gait and jump classification in modern equestrian sports. In Proceedings of the 2018 ACM International Symposium on Wearable Computers (pp. 88–91). ACM, 2018.

130. Marshall J. Smartphone sensing for distributed swim stroke coaching and research. Proceedings of the 2013 ACM conference on Pervasive and ubiquitous computing adjunct publication. ACM, 2013: 1413–1416.

131. Bächlin M, Förster K, Tröster G. SwimMaster: a wearable assistant for swimmer. Proceedings of the 11th international conference on Ubiquitous computing. ACM, 2009: 215–224.

132. Siirtola P, Laurinen P, Röning J, et al. Efficient accelerometer-based swimming exercise tracking. Computational Intelligence and Data Mining (CIDM), 2011 IEEE Symposium on. IEEE, 2011: 156–161.

133. Deng Z A, Hu Y, Yu J, et al. Extended Kalman Filter for Real Time Indoor Localization by Fusing WiFi and Smartphone Inertial Sensors. Micromachines, 2015, 6(4): 523–543.

134. Kon Y, Omae Y, Sakai K, et al. Toward Classification of Swimming Style by using Underwater Wireless Accelerometer Data. Ubicomp/ISWC'15 Adjunct, Osaka, Japan.

135. Woohyeok C, Jeungmin O, Taiwoo P, et al. MobyDick: An Interactive Multi-swimmer Exergame. Proceedings of the 12th ACM Conference on Embedded Network Sensor Systems, SenSys '14, Memphis, Tennessee, USA, November 3–6, 2014.

136. Anhua L, Jianzhong Z, Kai L, et al. An efficient outdoor localization method for smartphones. 23rd International Conference on Computer Communication and Networks, ICCCN 2014, Shanghai, China, August 4–7, 2014.

137. Kartik S, Minhui Z, Xiang Fa Guo, et al. Using Mobile Phone Barometer for Low-Power Transportation Context Detection. Proceedings of the 12th ACM Conference on Embedded Network Sensor Systems, SenSys '14, Memphis, Tennessee, USA, November 3–6, 2014.

138. Muralidharan K, Khan A J, Misra A, et al. Barometric phone sensors: more hype than hope! Proceedings of the 15th Workshop on Mobile Computing Systems and Applications. ACM, 2014: 12.

139. D. Xiao, Z. Yu, F. Yi, L. Wang, C. Tan & Guo, B. (2016). SmartSwim: An Infrastructure-Free Swimmer Localization System Based on Smartphone Sensors. In: Chang C., Chiari L., Cao Y., Jin H., Mokhtari M., Aloulou H. (eds) Inclusive Smart Cities and Digital Health. ICOST 2016. Lecture Notes in Computer Science, vol 9677. Springer, pp 222–234.

140. N. Duta. A survey of biometric technology based on hand shape. Pattern Recognition 42 (11), 2009, pp. 2797–2806.

141. K. Karu, and A.K. Jain. Fingerprint classification. Pattern Recognition 29 (3), 1996, pp. 389–404.

142. D. Malaspina, E. Coleman, R.R. Goetz, J Harkavy-Friedman, C. Corcoran, X. Amador, S. Yale, and J.M. Gorman Odor identification, eye tracking and deficit syndrome schizophrenia Biological Psychiatry 51 (10), 2002, pp. 809–815.

143. H.A. Park, and K.R. Park. Iris recognition based on score level fusion by using SVM. Pattern Recognition Letters 28 (15), 2007, pp. 2019–2028.

144. K Kurita. Human Identification from Walking Signal Based on Measurement of Current Generated by Electrostatic Induction. In Proceedings of the 2011 International Conference on Biometrics and Kansei Engineering (ICBAKE '11), 2011, pp. 232–237.

145. JJ Little, and JE Boyd. Recognizing People by Their Gait: The Shape of Motion. Journal of Computer Vision Research, 1998, 1(2): 1–32.

146. C. Nickel, C. Busch, S. Rangarajan, and M. Möbius. Using Hidden Markov Models for accelerometer-based biometric gait recognition. Signal Processing and its Applications (CSPA), 2011 IEEE 7th International Colloquium on, 2011, pp. 58–63.

147. D. Mulyono, and H.S. Jinn. A study of finger vein biometric for personal identification. International Symposium on Biometrics and Security Technologies, 2008, pp. 1–8.

148. Google ATAP. 2015. Welcome to Project Soli. Video. (29 May 2015.) Retrieved April 11, 2016 from https://www.youtube.com/watch?v=0QNiZfSsPc0.

149. G. Zanca, F. Zorzi, A. Zanella, and M. Zorzi. Experimental comparison of RSSI-based localization algorithms for indoor wireless sensor networks. Proceedings of the workshop on Real-world wireless sensor networks (REALWSN '08), 2008, pp. 1–5.

150. Y. F. Huang, T. Y. Yao and H. J. Yang. Performance of Hand Gesture Recognition Based on Received Signal Strength with Weighting Signaling in Wireless Communications. Network-Based Information Systems (NBiS), 2015 18th International Conference on, 2015, pp. 596–600.

151. W. Xi, J. Zhao, X. Y. Li, K. Zhao, S. Tang, X. Liu and Z. Jiang. Electronic frog eye: Counting crowd using Wi-Fi. IEEE INFOCOM. 2014, pp. 361–369.
152. Z. Yang, Z. Zhou, Y. Liu. From RSSI to CSI: Indoor localization via channel response. ACM Computing Surveys (CSUR), 2013, 46(2):25.
153. K. Ali, A. X. Liu, W. Wang, and M. Shahzad. Keystroke Recognition Using Wi-Fi Signals. In Proceedings of the 21st Annual International Conference on Mobile Computing and Networking (MobiCom '15), 2015, pp. 90–102.
154. C. Han, K. Wu, Y. Wang, and L. M Ni. Wifall: Device-free fall detection by wireless networks. In Proceedings of IEEE International Conference on Computer Communications (INFOCOM '14), 2014, pp. 271–279.
155. R. Nandakumar, B. Kellogg, and S. Gollakota. Wi-Fi gesture recognition on existing devices. Eprint Arxiv, 2014.
156. G Wang, Y Zou, Z Zhou, and K Wu. We Can Hear You with Wi-Fi! In Proceedings of the 20th Annual International Conference on Mobile Computing and Networking (MobiCom '14), 2014, pp. 593–604.
157. Y. Wang, J. Liu, Y. Chen, M. Gruteser, J. Yang, and H. Liu. E-eyes: device-free location-oriented activity identification using fine-grained Wi-Fi signatures. In Proceedings of the 20th annual international conference on Mobile computing and networking, 2014, pp. 617–628.
158. M. Shahzad, and S. Zhang. Augmenting User Identification with WiFi Based Gesture Recognition. Proceedings of the ACM on Interactive, Mobile, Wearable and Ubiquitous Technologies, 2018, 2(3), 134.
159. S. Palipana, D. Rojas, P. Agrawal, and D. Pesch. FallDeFi: Ubiquitous Fall Detection using Commodity Wi-Fi Devices, Proceedings of the ACM on Interactive, Mobile, Wearable and Ubiquitous Technologies, 2018, 1(4), 155.
160. R. Chellappa, C. Wilson, and S. Sirohev. Human and machine recognition of faces: a survey. Proceedings of IEEE vol. 83 (5), 1995, pp. 705–740.
161. D. Halperin, W. Hu, A. Sheth, and D. Wetherall. Tool release: Gathering 802.11n traces with channel state information. ACM SIGCOMM CCR 41(1):53.
162. S. Sen, J. Lee, K.-H. Kim, and P. Congdon. Avoiding multipath to revive inbuilding Wi-Fi localization. In Proceeding of ACM MobiSys, 2013, pp. 249–262
163. X. Li, S. Li, D. Zhang, J. Xiong, Y. Wang, and H. Mei. Dynamic-MUSIC: Accurate Device-free Indoor Localization. In Proceedings of the 2016 ACM International Joint Conference on Pervasive and Ubiquitous Computing (UbiComp'16). ACM, 2016, pp. 196–207.
164. X. Li, D. Zhang, Q. Lv, J. Xiong, S. Li, Y. Zhang, and H. Mei. IndoTrack: Device-Free Indoor Human Tracking with Commodity Wi-Fi, Proceedings of the ACM on Interactive, Mobile, Wearable and Ubiquitous Technologies, 2017, 1(3), 72.
165. D. Wu, D. Zhang, C. Xu, H. Wang, and X. Li. 2017. Device-Free WiFi Human Sensing: From Pattern-Based to Model-Based Approaches. IEEE Communications Magazine 55, 10 (OCTOBER 2017), 91–97.
166. D. Zhang, H. Wang, and D. Wu. Toward Centimeter-Scale Human Activity Sensing with Wi-Fi Signals. Computer 50, 1 (Jan 2017), 48–57.
167. H. Wang, D. Zhang, J. Ma, Y. Wang, Y. Wang, D. Wu, T. Gu, and B. Xie. Human Respiration Detection with Commodity WiFi Devices: Do User Location and Body Orientation Matter?. In Proceedings of the 2016 ACM International Joint Conference on Pervasive and Ubiquitous Computing, 2016, pp. 25–36.
168. D. Wu, D. Zhang, C. Xu, Y. Wang, and H. Wang. WiDir: Walking Direction Estimation Using Wireless Signals. In Proceedings of the 2016 ACM International Joint Conference on Pervasive and Ubiquitous Computing (UbiComp'16), 2016, pp. 351–362.
169. F. Zhang, D. Zhang, J. Xiong, H. Wang, K. Niu, B. Jin, and Y. Wang. From Fresnel Diffraction Model to Fine-grained Human Respiration Sensing with Commodity Wi-Fi Devices. IMWUT 2, 1 (2018), 53:1–53:23

170. T. Xin, B. Guo, Z. Wang, M. Li, Z. Yu and X. Zhou, "FreeSense: Indoor Human Identification with Wi-Fi Signals," 2016 IEEE Global Communications Conference (GLOBECOM), Washington, DC, 2016, pp. 1–7. DOI: https://doi.org/10.1109/GLOCOM.2016.7841847

171. Adib Fadel, Hongzi Mao, Zachary Kabelac, Dina Katabi, and Robert C Miller. 2015. Smart homes that monitor breathing and heart rate. In Proceedings of the 33rd Annual ACM Conference on Human Factors in Computing Systems. ACM, 837–846.

172. Healthcare. 2012. F. M. Market for embedded health monitoring-gadgets to hit 170M devices by 2017. http://www.fiercemobilehealthcare.com/story/market-embedded-health-monitoring-gadgets-hit-170m\-devices-2017/2012-08-03. (2012).

173. Michelle A Cretikos, Rinaldo Bellomo, Ken Hillman, Jack Chen, Simon Finfer and Arthas Flabouris. Respiratory rate: the neglected vital sign. Medical Journal of Australia 188, 11 (2008): 657.

174. J. Brian North and Sheila Jennett. Abnormal breathing patterns associated with acute brain damage. 1974. Archives of neurology 31, 5 (1974), 338–344.

175. Yuan, George, Nicole A. Drost, and R. Andrew McIvor. 2013. Respiratory rate and breathing pattern. McMaster University Medical Journal 10, 1, 23–25.

176. Cooke, Jana R., and Sonia Ancoli-Israel. Normal and abnormal sleep in the elderly. 2011. Handbook of clinical neurology/edited by PJ Vinken and GW Bruyn 98 (2011), 653.

177. Norman, Daniel, and José S. Loredo. Obstructive sleep apnea in older adults. Clinics in geriatric medicine 24.1 (2008), 151–165.

178. Lee-Chiong, L. Teofilo Monitoring respiration during sleep. Clinics in chest medicine 24, 2 (2003), 297–306.

179. R. Madeline, MPH. Vann. 2015. The 15 Most Common Health Concerns for Seniors. URL: http://goo.gl/EQn2fn, 2015

180. J. N. Wilkinson, and V. U. Thanawala. 2009. Thoracic impedance monitoring of respiratory rate during sedation – is it safe?. Anaesthesia, 64 (2009), 455–456.

181. Jaffe, B. Michael. 2008. Infrared measurement of carbon dioxide in the human breath: "breathe-through" devices from Tyndall to the present day. Anesthesia & Analgesia 107, 3 (2008), 890–904.

182. Rita Paradiso. 2003. Wearable health care system for vital signs monitoring. In Information Technology Applications in Biomedicine, 2003. 4th International IEEE EMBS Special Topic Conference on. IEEE, 283–286.

183. Shoko Nukaya, Toshihiro Shino, Yosuke Kurihara, Kajiro Watanabe, Hiroshi Tanaka. Non-invasive bed sensing of human biosignals via piezoceramic devices sandwiched between the floor and bed. IEEE Sensors journal 12, 3 (2012), 431–438.

184. Hulya Gokalp and Malcolm Clarke. Monitoring activities of daily living of the elderly and the potential for its use in telecare and telehealth: a review. TELEMEDICINE and e-HEALTH 19, 12 (2013), 910–923.

185. Jochen Penne, Christian Schaller, Joachim Hornegger, and Torsten Kuwert. Robust real-time 3D respiratory motion detection using time-of-flight cameras. International Journal of Computer Assisted Radiology and Surgery 3, 5 (2008), 427–431.

186. T Kondo, T Uhlig, P Pemberton, and PD Sly. Laser monitoring of chest wall displacement. European Respiratory Journal 10, 8 (1997), 1865–1869.

187. M Nowogrodzki, DD Mawhinney, and HF Milgazo. Non-invasive microwave instruments for the measurement of respiration and heart rates. NAECON 1984 1984 (1984), 958–960.

188. Svetha Venkatesh, Christopher R Anderson, Natalia V Rivera, and R Michael Buehrer. Implementation and analysis of respiration-rate estimation using impulse-based UWB. In Military Communications Conference, 2005. MILCOM 2005. IEEE. IEEE, 3314–3320.

189. Fadel Adib, Hongzi Mao, Zachary Kabelac, Dina Katabi, and Robert C Miller. 2015. Smart homes that monitor breathing and heart rate. In Proceedings of the 33rd Annual ACM Conference on Human Factors in Computing Systems. ACM, 837–846.

190. Ruth Ravichandran, Elliot Saba, Ke-Yu Chen, Mayank Goel, Sidhant Gupta, and Shwetak N Patel. 2015. WiBreathe: Estimating respiration rate using wireless signals in natural settings in the home. In Pervasive Computing and Communications (PerCom), 2015 IEEE International Conference on. IEEE, 131–139.
191. Fadel Adib, Zachary Kabelac, Dina Katabi, Robert C. Miller. 2013. 3D Tracking via Body Radio Reflections. In Proceedings of the 11th USENIX Conference on Networked Systems Design and Implementation. USENIX Association, 317–329.
192. Hao Wang, Daqing Zhang, Junyi Ma, Yasha Wang, Yuxiang Wang, Dan WU, Tao Gu, Bing Xie. 2016. Human respiration detection with commodity WiFi devices: do user location and body orientation matter?. In Proceedings of the 2016 ACM International Joint Conference on Pervasive and Ubiquitous Computing. ACM, 25–36.
193. Jian Liu, Yan Wang, Yingying Chen, Jie Yang, Xu Chen, and Jerry Cheng. 2015. Tracking Vital Signs During Sleep Leveraging Off-the-shelf WiFi. In Proceedings of the 16th ACM International Symposium on Mobile Ad Hoc Networking and Computing. ACM, 267–276.
194. Xuefeng Liu, Jiannong Cao, Shaojie Tang, and Jiaqi Wen. 2014. Wi-Sleep: Contactless sleep monitoring via WiFi signals. In Real-Time Systems Symposium (RTSS), 2014 IEEE. IEEE, 346–355.
195. Xuefeng Liu, Jiannong Cao, Shaojie Tang, Jiaqi Wen, and Peng Guo. 2016a. Contactless Respiration Monitoring via WiFi Signals. Mobile Computing, IEEE Transactions on (2016).
196. Chenshu Wu, Zheng Yang, Zimu Zhou, Xuefeng Liu, Yunhao Liu, and Jiannong Cao. Non-Invasive Detection of Moving and Stationary Human With WiFi. Selected Areas in Communications, IEEE Journal on 33, 11 (2015), 2329–2342.
197. Heba Abdelnasser, Khaled A Harras, and Moustafa Youssef. 2015. Ubibreathe: A ubiquitous non-invasive WiFi-based breathing estimator. arXiv preprint arXiv:1505.02388 (2015).
198. Ossi Kaltiokallio, Huseyin Yigitler, Riku Jantti, and Neal Patwari. 2014. Non-invasive respiration rate monitoring using a single COTS TX-RX pair. In Information Processing in Sensor Networks, IPSN-14 Proceedings of the 13th International Symposium on. IEEE, 59–69.
199. Neal Patwari, Lara Brewer, Quinn Tate, Ossi Kaltiokallio, and Maurizio Bocca. 2014a. Breathfinding: A wireless network that monitors and locates breathing in a home. Selected Topics in Signal Processing, IEEE Journal of 8, 1 (2014), 30–42.
200. Neal Patwari, James Wilson, Sundaram Ananthanarayanan, Sneha Kumar Kasera, and Dwayne R Westenskow. 2014b. Monitoring breathing via signal strength in wireless networks. Mobile Computing, IEEE Transactions on 13, 8 (2014), 1774–1786.
201. Philippe Arlotto, Michel Grimaldi, Roomila Naeck and Jean-Marc Ginoux. An ultrasonic contactless sensor for breathing monitoring. Sensors 14.8 (2014), 15371–86.
202. Rajalakshmi Nandakumar, Shyamnath Gollakota, Nathaniel Watson M.D.. 2015. Contactless Sleep Apnea Detection on Smartphones. In Proceedings of the 13th Annual International Conference on Mobile Systems, Applications, and Services. ACM, 45–57.
203. Carina Barbosa Pereira, Xinchi Yu, Michael Czaplik, Rolf Rossaint, Vladimir Blazek, and Steffen Leonhardt. Remote monitoring of breathing dynamics using infrared thermography. Biomedical optics express, 6, 11 (2015), 4378–4394.
204. Hao Tian, Guoliang Xing, and Gang Zhou. 2013. iSleep: unobtrusive sleep quality monitoring using smartphones. In Proceedings of the 11th ACM Conference on Embedded Networked Sensor Systems. ACM, 1–14.
205. Chunyi Peng, Guobin Shen, Yongguang Zhang, Yanlin Li, Kun Tan (2007). BeepBeep: a high accuracy acoustic ranging system using COTS mobile devices. In Proceedings of the 5th international conference on Embedded networked sensor systems. ACM, 1–14
206. Tian Hao, Guoliang Xing, Gang Zhou. 2015. RunBuddy: A Smartphone system for running rhythm monitoring. In Proceedings of the 2015 ACM International Joint Conference on Pervasive and Ubiquitous Computing. ACM, 133–144

207. A. G. Stove. (1992). Linear FMCW radar techniques. Radar & Signal Processing Iee Proceedings F, 139, 5(1992):343–350.
208. Oppenheim A V, Willsky A S, Nawab S H. 1996. Signals & systems (2nd ed.) Prentice-Hall, Inc.
209. S. Suleymanov. (2016). Design and Implementation of an FMCW Radar Signal Processing Module for Automotive Applications (Master's thesis, University of Twente).
210. T. Wang, D. Zhang, Y. Zheng, T. Gu, X. Zhou, B. Dorizzi. C-FMCW Based Contactless Respiration Detection Using Acoustic Signal. Proceedings of the ACM on Interactive, Mobile, Wearable and Ubiquitous Technologies table of contents archive, 1(4), December 2017, Article No. 170

Chapter 6
Group Behavior Recognition

Abstract Compared with individual behavior sensing and understanding, the recognition of group behavior is more challenging. In this chapter, we present some of our recent progress on group behavior sensing and recognition. Specifically, in Sect. 6.1, we discuss how to recognize the mobility level and structure of groups in the physical world by leveraging mobile devices. Afterwards, we present the recognition of human semantic interactions and group interaction patterns in smart spaces in Sects. 6.2 and 6.3. Finally, we discuss how we can organize, suggest, or predict group activities by leveraging the power of mobile crowdsensing and mobile social networking in Sects. 6.4 and 6.5.

6.1 Recognition of Group Mobility Level and Group Structure with Mobile Devices[1]

6.1.1 Introduction

Proximity-based groups exist in many scenarios, such as an opportunistic group in a shopping mall [1], or a team in a meeting room [2]. People in these groups are close to each other [2, 3] or have similar trajectories [1, 4–6]. Different from a large community, such a proximity-based group is usually formed by a relatively small number of people and can have a particular shape. The group members may occasionally or regularly interact with each other. Monitoring the dynamics (e.g., shapes, mobility levels, and interactions among group members) of such a group can facilitate a set of services, e.g., enhancing the members' social relationships and providing group-aware recommendations. Therefore, we aim to understand such group dynamics in this work.

---

[1]Part of this section is based on a prevous work: H. Du, Z. Yu, F. Yi, Z. Wang, Q. Han and B. Guo, "Recognition of Group Mobility Level and Group Structure with Mobile Devices," in IEEE Transactions on Mobile Computing, vol. 17, no. 4, pp. 884–897, 1 April 2018. DOI: https://doi.org/10.1109/TMC.2017.2694839

In particular, we mainly focus on two typical group dynamics, i.e., group shape and group mobility level. Group mobility level can be stationary, strolling, running, etc.; group shape or structure can be a queue, an irregular polygon, or a straight line side by side. Combining the two basic characteristics, we can describe mobile group activities with more details, such as strolling or stationary side by side, and walking or running in a queue. The mobility patterns and structures of groups may be distinct in different scenarios. For example, a shopping group of two or three people usually walks in parallel with a low speed, whereas people in an indoor recreation center may run in a single line along the running track. Moreover, the group structure is also important for social relationship identification. In particular, a group structure can be described based on the members' positions, which include the front, back, left, or right. The formation of group structure may imply how members communicate with each other [7]. For instance, walking side by side is more convenient for communications [7]. In addition, members' positions might correlate with certain social status in some scenarios. For example, in the case of the front-back relation, the leaders or directors often walk in front, followed by the staff. In the case of the left-right relation, the active person who talks more is likely to walk in the middle of the group and the others walk on the side [8]. Therefore, to fully understand the social relations and status among a group, we need to obtain not only the front-back relation (i.e., leader-follower) but also the left-right relation. Accordingly, monitoring group mobility and structure is crucial for understanding group activities and social relationships.

In recent years, lots of efforts have been focused on group dynamics, e.g., group detection, intragroup relations, and group behaviors [1, 4, 9–14]. Prior work has emphasized the need for a better understanding of group dynamics and only touched on group mobility patterns and the leadership relation in a group. Some works investigated the walking behaviors of a group, where the group structure was proposed implicitly. Utilizing simulation, Moussaid et al. [8] investigated the spatial group organizations. Costa [7] studied the interpersonal distance in group walking based on video image. However, these methods cannot achieve group monitoring in near real time, and image-based algorithm needs large storage. Therefore, a novel technology is required. Fortunately, advanced sensing technologies enabled wide adoption of mobile devices in context-aware applications [15–21], so it is possible for group monitoring to be done in a ubiquitous and nonintrusive way. With smartphones as powerful sensors, most of the existing studies mainly study the detection of different groups, e.g., detection of shopping groups using Wi-Fi, compass, accelerometer, and barometer readings [1], or detection of meeting groups using Bluetooth signals [2]. However, we have limited knowledge on group structures or fine-grained group mobility level. Specifically, in some cases, knowing whether a group of people is moving or stationary is enough [1, 9], whereas in some cases we need to know the moving speed, i.e., lower speed strolling, normal-speed walking, or higher speed running, which can potentially facilitate group activity understanding. Moreover, the concrete structure of a moving group, e.g., in parallel, in a straight line, or in an irregular polygon, is another useful piece of information that can be used to infer group members' social relations. Therefore, we aim to utilize pervasively deployed Wi-Fi combined with sensors already built in most commodity smartphones to classify group mobility and recognize group structures.

Recognizing group mobility level and group structure poses several interesting challenges. First, group mobility level recognition requires the consideration of the mobility of all group members. Different from individual mobility detection, recognizing group mobility level focuses on the entirety of a group, for which the object to be classified is the data combined from all group members. Second, group structure recognition is an issue involving each member's position in the same group. Accurate localization has been a difficult problem. Due to privacy, precision, or energy consumption concerns, Bluetooth or GPS-based positioning is not suitable for structure recognition of small groups with 2–4 members. Therefore, we decided to explore the solution for structure recognition without knowledge of absolute positions. To our delight, there is no need to localize each member as our problem only requires knowing the relative position relationship. Existing research has proposed identifying the leadership in a group [9, 22], i.e., who is in the front and who is behind. However, to fully understand the group structure, the left-right relation recognition is also important, which has been little studied in existing work. Therefore, recognizing the group structure by using mobile sensing is another challenge.

In this chapter, we propose algorithms for fine-grained mobility level classification and structure recognition of social groups utilizing mobile devices [23]. Specifically, we first detect group mobility behaviors by utilizing the accelerometer-embedded in smartphones, and then apply a supervised learning algorithm to recognize four levels of group mobility, i.e., stationary, strolling, walking, and running. Second, we propose a method that uses relative position estimation to recognize the structures of moving groups by leveraging multiple types of mobile sensor data, where the position relations among group members are described as the front (i.e., leader), back (i.e., follower), left, and right. Third, for the left-right position recognition we propose interaction discovery utilizing acoustic information captured by smartphones.

6.1.2 Related Work

6.1.2.1 Detecting Groups

The origin of groups can be different, not only pedestrian flocks [4–6], but also maybe a queue waiting for the service, a coincidence companion in a shopping mall or a team discussing in a meeting room. Detecting such a group is of great importance for better information sharing and service providing. QueueSense [3] is a queue management system, which measures the disparity of people in different lines and partitions them into different queues. Grace [2] studied Bluetooth signal strength and focused on recognizing groups where people have face-to-face interactions. Both of them need a relatively high accuracy in distance estimation, which is not required in our work. Besides, Yu et al. [24] proposed to identify subgroups in a homogeneous activity group. They applied a two-stage process to identify sub-groups. They also proposed to discover context-aware community groups leveraging proximity-based mobile networks [25]. Gordon et al. [26] presented a method for distributed, peer-to-peer recognition of group affiliations in multigroup environments, using the divergence of mobile phone sensor data.

6.1.2.2 Recognizing Group Behaviors

Research on group behavior recognition covers many scenarios in reality [21, 27–29]. Feese et al. [21] utilized smartphones and detected group proximity dynamics of firefighting teams. Based on sensor signals and social signal processing, Dim et al. [27] detected typical social behavior patterns of visitor pairs, knowing the social context. Utilizing deployed Wi-Fi infrastructure and accelerometer from smartphones, Stisen et al. [29] recognized task phase for mobile workers in large buildings, such as in a hospital. Martella et al. [30] explored the usefulness of proximity sensor to mine the visitors' behavior in museums; their implementation shows that group behavior can be recognized by means of data clustering and visualization. Applications used in these works may be built on our framework.

On the other hand, the leadership relation among the group members is a significant group behavior, which is also part of our work. Being a leader or a follower reflects the role of the member in his/her group, and generally has his/her own nonnegligible effect on the group. The recognition of such relations could be inspired by leader discovery from correlated time series [31, 32], which provides basic knowledge for our leadership recognition and has been used in market research, reality mining, etc. Recently, several studies of identifying following and leadership behavior of pedestrians have been conducted. Sanchez-Cortes et al. [22] studied how to identify emergent leaders in small groups utilizing visual features. Similarly, Kjærgaard et al. [9] detected following and leadership behavior of pedestrians based on mobile sensing data (e.g., accelerometer, gyroscope data). Further, Beyan et al. [33] proposed to identify leadership in small group meetings leveraging the power of multitask learning. They applied nonverbal audiovisual features and the results of social psychology questionnaires in a joint learning manner, and achieved better performance compared to single-task learning methods.

Based on the leadership relation, we propose to recognize the group structure utilizing the relative position relations. The principle for the leader-follower recognition in Kjærgaard et al. [9] is partly adopted as a solution, but not the main work in this chapter. Moreover, we have an improvement and simplification compared to leadership detection [9]. In addition, we aim to recognize the group structure, where not only the leader-follower relation needs to be known, but also the left-right relation. Specifically, in our ideas on leadership relation recognition, we define a concept of direct leader and propose a simplified PageRank algorithm for clearly recognizing the global group position relations. We further consider the left-right relations, which have rarely been studied before. In our prior work [23], we detected leader-follower relation utilizing cross-correlation, and recognized left-right relation based on orientation change caused by interaction. In this chapter, we address the limitations in our prior method on left-right relation recognition by proposing interaction discovery utilizing acoustic information captured by smartphones. For leadership relation recognition, we also added another similarity metric, i.e., DTW, and compared its effectiveness with the existing one [23]. More importantly, unscripted experiments are conducted and evaluated in this work. Furthermore, we developed the proposed framework and presented two prototype applications as well as a user study.

6.1.2.3 Monitoring Crowd Dynamics

The crowd can be regarded as a set of groups and the study of crowd behaviors is of great significance not only in urban computing [34, 35] but also in some major events. In order to manage the mobility of crowds as a controlled process, Blanke et al. [36] designed an application which offered information about a festival allowing to collect positions of the visitors continuously. Similarly, to analyze the complex spatiotemporal dynamics of crowd movements at mass events, Versichele et al. [37] proposed a proximity-based Bluetooth tracking mechanism. As to city-scale applications for crowd behaviors, Wirz et al. [38] explored location-aware smartphones to monitor crowds, and inferred the crowd density based on the walking speed of pedestrians.

Due to the inherent nature of group dynamics, we focus on monitoring changes inside of groups. In this chapter, we propose a fine-grained mobility classification and structure recognition approach for social groups utilizing mobile devices. Leveraging the potential ability of built-in sensors on smartphones, we can precisely detect the group moving pattern and the relative position relations among members in the same group for structure recognition.

6.1.3 System Overview

Figure 6.1 shows the two major components of the proposed approach, i.e., the group mobility classification module and the group structure recognition module, along with the data capture module, which collects and prepares data for the two major components.

Data Capture: We collect the acceleration and magnetic field data with the phone in three directions: horizontal, vertical, and pointing to the ground. In addition, we collect Wi-Fi signal strength data, which provides localization information for the leader-follower recognition. We also collect audio wave data via a microphone on the phone, which is then used to discover interaction among group members and further help recognize the left-right relation. The collected data has to be preprocessed to eliminate noise and spikes before being sent to the two major

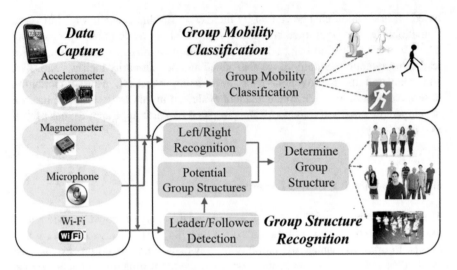

Fig. 6.1 Overview of the proposed approach

modules. During the preprocessing, a low-pass filter is used to eliminate the high-frequency part.

Mobility Classification: This module recognizes four levels of group mobility, i.e., stationary, strolling, walking, and running. A supervised learning technique is applied to model the features extracted from the accelerometer sensor data.

Structure Recognition: The structure of a group depends on each member's position. Instead of accurate positioning, we only need to obtain relative position relations among group members. By fusing data from the accelerometer, magnetic field, microphone, and Wi-Fi, this module recognizes the pair-wise relative location: the leader, the follower, the left, and the right. Furthermore, the leader-follower relations help recognize several potential group structures, and the left-right relations help differentiate members side by side and determine the final group structure from the several potential structures.

6.1.4   Group Mobility Classification

Mobility patterns, e.g., strolling, walking, or running, can be regarded as pedestrians' natural paces. Different group mobility patterns might indicate different group contexts. For example, in a public place, walking is a typical mobility pattern and an even lower moving speed may be common in a shopping mall (i.e., strolling). In addition, we also consider running as a mobility pattern, which is faster than walking and might be an indication of a latent danger caused by anomaly events or chaos in public places. What distinguishes our work from others [1, 4, 9] is that we classify group mobility in a more fine-grained manner: stationary, strolling, walking, and running.

In general, people in the same group have similar acceleration. The accelerometer sensor data from mobile phone reflects users' moving acceleration, to some extent. Therefore, group mobility identification can be realized by differentiating users' acceleration. Compared with existing studies, this work aims to identify more fine-grained group mobility levels. In particular, individual acceleration is a main feature in group detection that helps identify people with similar trajectories, whereas group mobility classification considers all members in the same group. Group mobility cannot be identified solely based on the feature of a single member, but the collective motion features of all the members. Therefore, the training data for each moving pattern is a continuous data wave obtained from each group member's acceleration data. Specifically, waveforms of all the members in the same group are concatenated, forming a single waveform for feature extraction. Figure 6.2 shows a part of the waveform collected from three members in the same group when the group has four mobility levels.

In this chapter, we apply a supervised learning algorithm to classify the data stream into four levels corresponding to four mobility patterns: stationary, strolling, walking, and running. Each data record is a time series of acceleration metadata sampled at 100 Hz. In the data preprocessing stage, a low-pass filter is used to eliminate the high-frequency component so that possible noise and spikes are removed. We then use relative data instead of absolute acceleration magnitude to improve the accuracy during the feature extraction stage. The reason is that the acceleration data of different members in the same group may vary significantly; for example, users are likely to put phones in different positions, and the embedded sensors of heterogeneous devices may have inherent errors [39–42]. Assuming that the acceleration change in a group is generally synchronous, the standard deviations of acceleration metadata are extracted as var_x, var_y, and var_z, with x, y, and z representing acceleration data in three directions. Moreover, $var._{\triangle x}$, $var._{\triangle y}$, and $var._{\triangle z}$ are also extracted as features, which are the corresponding standard deviation of difference between adjacent data. With the extracted features, a classifier is applied for the four-level mobility classification.

6.1.5 Group Structure Recognition

Intuitively, the group structure can be reflected by the relative position relations among multiple members in a group. In particular, we are interested in identifying the leader-follower and the left-right relationships between any two group members. The leader-follower relation helps recognize possible group structures with the same direct leader relations, and then the left-right relation further determines the final unique structure from all the possible structures by differentiating parallel members.

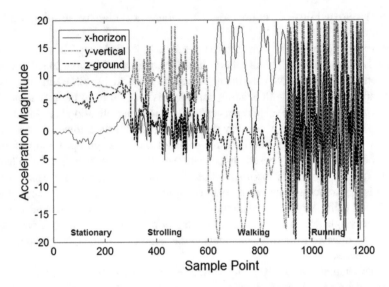

Fig. 6.2 Acceleration waves when the group is moving with different mobility levels

6.1.5.1   Leader-Follower Recognition

The leader or follower means a member walking in front or behind and the relations help recognize the basic group structure, i.e., formation of a queue versus in a row, etc.

We would like to determine direct leaders from a set of time-series data by analyzing lead-lag relations. Members in the same group tend to share similar trajectory but the time of arriving at a certain location is different. The leader arrives first, whereas the follower arrives later with a time lag d. The follower and leader still exhibit highly similar time series with the difference d between start time. The time-lagged method [9] is adopted for relation recognition with a different similarity metric. Further, we propose a new algorithm to identify the direct leader.

Our time-series data is composed of Wi-Fi signals and accelerometer sensor data in a time window t, represented as an l-dimension series $S_i = \{s_i^1, s_i^2, \ldots, s_i^l\}$, which reflects the mobility trajectory of a member i, and whose corresponding average value is described as p_i. Then a set of time series $\{S_1, S_2, .., S_n\}$ is used to represent all the members' moving trajectories in a group. To measure time-lag similarity, we utilize two metrics in our work, cross-correlation [43] and DTW, which outperform other metrics for leadership detection as demonstrated in Kjærgaard et al. [9]. We will analyze and compare their performance in the environment, and then choose the better one. We use cross-correlation and DTW to measure the similarity between S_i and S_j with lag d, which extends typical similarity metrics with a time lag, shown as follows:

$$r_{i,j}(d) = \frac{\sum_{k=1}^{n}\left[\left(s_i^k - p_i\right)\left(s_j^{(k-d)} - p_j\right)\right]}{\sqrt{\sum_k \left(s_i^k - p_i\right)^2}\sqrt{\sum_k \left(s_j^{(k-d)} - p_j\right)^2}}, \tag{6.1}$$

$$r_{i,j}(d) = \frac{1}{dtw\left(S_i(k), S_j(k-d)\right)}. \tag{6.2}$$

When the similarity coefficient exceeds a certain threshold, the two series will be assigned to the same class. The extended cross-correlation or DTW analyzes the relation between series S_i with beginning time k, and S_j with beginning time $k - d$, described as $S_i(k)$ and $S_j(k - d)$. In other words, this metric compares one user's trajectory starting from time k and another user's trajectory starting from time $k - d$. If the obtained similarity coefficient is large enough, it means that the routes of these two users are similar with a time lag d. If d is positive, the second user arrives at each position later than the first one. If d is close to zero, $r_{i,j}(d)$ is the original cross-correlation or DTW, and that means the two users arrive at most of the positions at the same time; that is, they are in parallel. Otherwise, the second user arrives earlier than the first one.

The recognition of leader-follower should consider different time lag d. Specifically, we use the lag q_{est} defined as follows:

$$q_{est} = \frac{1}{sum}\sum_{-m \le d \le m}\frac{1}{r_{i,j}(d)} \times d. \tag{6.3}$$

If q_{est} is negative, the second user is the follower; if q_{est} is close to zero, then the two users are in parallel; otherwise, the second one is the leader. Finally, in our work, a relation matrix recording the corresponding values of q_{est} is obtained. By removing duplicate relations, the matrix can be represented as an upper triangular matrix, $H = [h_{i,j}]_{n \times n}$, as shown below, where $i < j$, and $h_{i,j} = q_{est}$:

$$H = \begin{bmatrix} 0 & h_{1,2} & h_{1,3} & \cdots & h_{1,n} \\ 0 & 0 & h_{2,3} & \cdots & h_{2,n} \\ \vdots & \vdots & \vdots & \ddots & \cdots \\ 0 & 0 & 0 & \cdots & h_{n-1,n} \\ 0 & 0 & 0 & \cdots & 0 \end{bmatrix}. \tag{6.4}$$

With the relation matrix, a set of pairs indicating the relations between any two group members can be obtained. To recognize group structure, the global relation and the direct leader should be identified. To this end, leadership transitivity is utilized to find the global leader-follower order. In other words, leadership transitivity can tell that a user is the leader of all users walking behind him/her, not just the leader of the one who is following him/her in the nearest position. The more followers a leader has, the more front his/her position is in the group.

For the direct leader recognition, we propose a simplified PageRank method, named "GScore," which is an improvement and simplification compared to Kjærgaard et al. [9]. It scores every time series and identifies the order of group members via ranking their scores. Besides the transitivity, PageRank algorithm scores every page with the weight, whereas we only want to recognize the leader and do not need to compute the accurate scores. Therefore, GScore only needs to guarantee that users with more followers get higher scores, for which the leadership transitivity involved is enough. Without the weight calculation, its complexity is much lower than the PageRank algorithm. We first give each time series a basic score $score_0$, and set a score slice t. Then, update the score of a time series S_i, which is determined by the relations between S_i and S_{i+k}. If user$_{i+k}$ is the leader of user$_i$, update user$_i$'s score by reducing a score slice t and add a slice to user$_{i+k}$'s score at the same time. If user$_{i+k}$ is the follower of user$_i$, add a slice to user$_i$'s score and reduce a slice from the score of user$_{i+k}$. Once user$_{i+k}$ is the last one in the time series, i.e., $i + k = n$, it would be the last updating for user$_i$'s score. GScore repeats the above steps to complete updating the scores of user$_{i+1}$, user$_{i+2}$, ..., and user$_n$. The process eventually confirms each user's score by retrieving all relation elements in relation matrix H, where the score of user$_i$ is calculated as follows:

$$\begin{cases} \text{score}(i)=\text{score}_i+\tau \times g\left(h_{i,j}\right) j>i, \\ \text{score}(j)=\text{score}_j-\tau \times g\left(h_{i,j}\right) j>i. \end{cases} \text{where } g\left(h_{i,j}\right) = \begin{cases} 1 & h_{i,j} < 0, \\ -1 & h_{i,j} > 0. \end{cases} \tag{6.5}$$

An instance demonstrating the running process of the algorithm is illustrated in Fig. 6.3.

Finally, once all the scores have been updated, sort them in a descending order and obtain a sequential user series U, where users with a higher score would take a more forward position, i.e., have more followers. We define one's nearest leader in geography as his/her direct leader that can be easily obtained from series U. Specifically, if a follower user$_i$ has two parallel leaders, e.g., user$_j$ and user$_k$, the one who is more similar to user$_i$ is defined as his/her direct leader.

Based on the time-lag method and the GScore algorithm, the direct leaders of all group members can be recognized, and then we can obtain the basic position relations in a group. Figure 6.3 shows the structures of two groups and the results of employing the GScore algorithm, where solid lines represent the direct leader relations between members in a group and the positive scores or negative scores are presented in dashed lines.

6.1.5.2 Left-Right Recognition

As shown in Fig. 6.3, given a pair of parallel users (e.g., user$_1$ and user$_2$, or user$_3$ and user$_4$), it is still necessary to determine who is on the left or right to obtain a clear

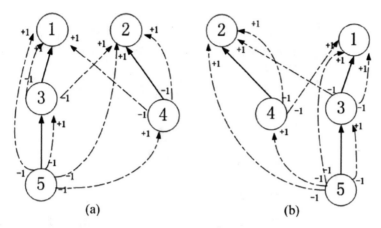

Fig. 6.3 Group structures with the direct leader relations and adding/reducing score relations (solid lines denote the direct leader-follower relations; dashed lines show adding score or reducing score links in GScore algorithm)

group structure. The left-right relations reflect social relations between parallel members and may indicate social status in some culture. We propose an approach to distinguish the left-right relation based on social interactions captured by mobile sensing data.

Usually people walking in parallel may have interactions with each other. For instance, a person on the left would slightly turn his/her body right from time to time for better interaction with his/her companion and vice versa. When there is no interaction, the parallel members' body directions would have similar angle or a small stable difference. Otherwise, the difference in two members' body orientations would be much larger, more irregular, and unstable when they have interactions. Therefore, we argue that the left-right relation can be recognized by comparing the difference between parallel members' body directions with or without interaction.

Determination of body orientation or heading direction is important yet difficult. The inertial sensors built in smartphones can only detect the orientation of the phone itself. However, the phone orientation varies during walking, and it may be different from the user orientation, with a phone heading offset. Therefore, reliable user direction inference is an extremely difficult task. To the best of our knowledge, there has been no reliable solution to this problem yet.

In this part, we consider the case where the phone is put in the user's pants pocket. If phones are at the same position for parallel members, it can be argued that from the difference in parallel members' phone directions, the left-right relation can be recognized. This approach does not need the initial phone heading offset; that is, the user's heading direction is not necessary, which is different from accurate localization [44]. Figure 6.4 illustrates the orientation waves of two parallel members, where the differences in orientations are relatively stable without interaction, whereas there are apparently different variations in fluctuation during an interaction.

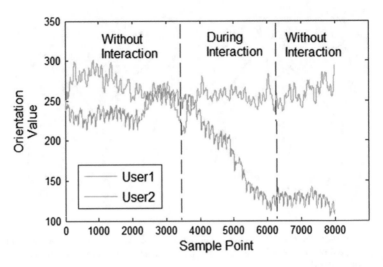

Fig. 6.4 Orientations of parallel members with and without interaction

Therefore, by comparing the difference between two parallel users' phone directions with or without interaction, the relative position can be recognized.

At first, the direction or orientation of a user's phone is obtained utilizing acceleration and magnetic field data collected from user's phone sensors. It is a common knowledge that the phone's orientation fluctuates when the user is walking and has no interaction with another user. When there are interactions, one or both of the parallel members' orientations will have an abrupt significant change. It then remains stable for a few seconds, e.g., 3–4 s, followed by another change and finally goes back to its original place. We regard an interaction as a non-instantaneous process. Therefore, we next compare the difference of each pair-wise parallel members' orientations in a time window T, and record the fluctuation range of every sliding window Δt. For a sliding window, we focus on points where one or both of parallel members have a sudden change in orientation, and regard such a point as an interaction point. Then, by conducting multiple pairs of comparable experiments with and without interaction, an empirical threshold for recognizing the interaction-based orientation change is obtained. If the fluctuation range exceeds the threshold, there may exist an orientation change caused by interaction. The interaction-based orientation change discovery mechanism is presented as follows:

$$\text{OC}_{\text{inter}}\left(o_i(t,t+\Delta t),o_j(t,t+\Delta t)\right) = \begin{cases} 0, & \text{if } o_i(t,t+\Delta t) \leq \theta \text{ and } o_j(t,t+\Delta t) \leq \theta \\ 1, & \text{otherwise} \end{cases}.$$

$$(6.6)$$

where the function $\text{OC}_{\text{inter}}(o_i, o_j)$ describes the relation between user$_i$ and user$_j$, and its value is 1 if interaction-based orientation change happens. $o_i(t, t + \Delta t)$ represents the orientation range of user$_i$'s phone from time t to $t + \Delta t$; t_0 is the start time of time window T; and θ is an empirical threshold.

Furthermore, to ensure the accuracy in left-right position detection, it is important to determine if the change in orientation is caused by face-to-face or back-to-back interaction, since true interaction exists only in face-to-face cases. If changes in orientation are detected, we use the microphone embedded in a smartphone to capture audio wave and discover the actual interaction. Groups often exist in noisy environments, so we need to distinguish group communication from noise in an audio wave. Ambient noise can be recognized by root mean square (RMS) and spectral entropy [45, 46]. RMS of ambient noise usually has a smaller value than a predefined threshold and entropy is higher. Therefore, using a supervised learning algorithm, we can acquire the two thresholds and recognize the actual interaction as well as the orientation change caused by interaction. This interaction detection method can be described as follows:

$$\text{InterOC} = \text{True} \Leftarrow \begin{cases} \text{OC}_{\text{inter}}\left(o_i, o_j\right) = 1, \\ \text{satisfy}(\text{rms}, \text{th}_{\text{rms}}) = 1, \\ \text{satisfy}(\text{se}, \text{th}_{\text{se}}) = 1. \end{cases} \tag{6.7}$$

where rms and se are the RMS and spectral entropy of the audio, respectively, and th_{rms} and th_{se} are the thresholds. The orientation change is caused by face-to-face interaction, i.e., InterOC = Ture, when rms and se satisfy th_{rms} and th_{se}, respectively. Specifically, satisfy(rms, th_{rms}) and satisfy(se, th_{se}) are used to determine whether the RMS and spectral entropy of the audio satisfy the threshold, respectively.

Finally, comparing with the phone's direction when there is no interaction, we find that a sudden increase of the orientation values generally means turning right, and decrease indicates turning left. Also note that the orientation data from acceleration sensor and magnetic field sensor ranges from 0 to 360, where 0 represents north, 90 represents east, 180 represents south, and 270 represents west. Specifically, if the original orientation is close to zero, the value increases regardless of whether a user turns right or left, which can be distinguished as follows. With the knowledge of orientation and value, the rise to 30–90 is a turn towards east (i.e., right), and the rise to 270–360 means turning towards west (i.e., left).

In short, we recognize true interaction for parallel members based on the captured audio signal and discover interaction-based orientation change using a sliding window, from which the left-right relation is derived. This approach for orientation change discovery does not need to obtain the initial relation between phones and users. We simply leverage pair-wise parallel members' phone orientations and orientation fluctuations during the sliding window.

6.1.5.3 Group Structure Determination

In the previous sections, we already recognize the leader-follower and the left-right relations in a group, and obtain the global relative position relation.

Utilizing recognized parallel pairs and direct leader relations, a group can be described as some structures that have the same direct leaders deterministically. But it is still not clear how the other members are relative to each other in positions. In combination with the left-right relations of the parallel members, we can further determine who is in front and who is on the left, and then finalize the proper group structure.

6.1.6 Performance Evaluation

To evaluate the performance of the proposed system, we developed a data collection application, named GCollector, for commodity smartphones running Android. GCollector gathers accelerometer, magnetometer, acoustic, and Wi-Fi connection recordings from the phone. Experiments were conducted in an office building, a campus, and a shopping mall, respectively, and different scenes were designed to evaluate different parts of the proposed system, as shown in Table 6.1. For more details about experimental setup and data collection, please refer to the article [47].

6.1.6.1 Experimental Results of Mobility Classification

For data collected during any 3 s, we extracted a set of features, including $var._x$, $var._y$, $var._z$, $var._{\Delta x}$, $var._{\Delta y}$, and $var._{\Delta z}$, and adopted the average precision and average recall as metrics to measure the performance of different classifiers. In particular, we applied several classification models to classify the mobility level, including decision tree, K-nearest neighbor (KNN), naive Bayes (NB), and SVM, as shown in Fig. 6.5.

According to Fig. 6.5a, we find that KNN achieves the best performance for mobility classification with a precision up to 99.5% and recall of 99.5%, while the precision of decision tree, naive Bayes, and SVM is 98.5, 97.2, and 97%. In addition,

Table 6.1 Scenes of the experiments

Module name	Description
Mobility level	3 groups with 4 moving patterns in the office building and campus with free behaviors
Scripted structure	3 groups walking in a straight line, parallel, and an irregular polygon in the office building
Unscripted structure	5 groups walking with free behaviors in the office building and shopping mall

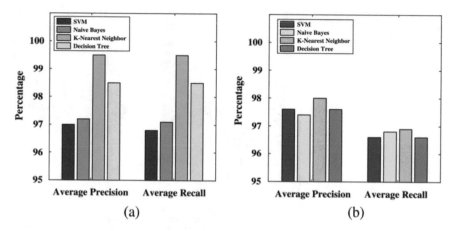

Fig. 6.5 Precision and recall of different classifiers. (**a**) Results based on tenfold cross-validation; (**b**) results based on leave-one-group-out cross-validation

as the tenfold cross validation may lead to overfitting [39], we also used the leave-one-group-out cross-validation for evaluation. As illustrated in Fig. 6.5b, the four models generated similar results, and KNN still achieved the best performance.

6.1.6.2   Experimental Results of Structure Recognition

The performance of structure recognition is measured by the accuracy of the leader-follower recognition and the left-right recognition.

First, based on Wi-Fi information and acceleration data, we adopted cross-correlation and DTW, respectively, to measure lag similarity and obtain the lag q_{est}. After analyzing the results of these two methods, leader-follower relation identification is evaluated for both scripted and unscripted experiments. It is observed that in scripted experiments, using a time window of 6 s, the accuracy of cross-correlation is up to 80%, which is higher than using DTW. In unscripted experiments, DTW achieves an accuracy of 72.7% in a shopping mall and 66.7% in an office building, which is similar to the performance of cross-correlation. However, we find that the DTW-based method is much more stable, and thus use it to detect leader-follower relation for unscripted experiments.

Second, the left-right recognition is based on orientation changing detection (i.e., interaction detection) of the phones carried by parallel subjects. In the experiments, we discovered interactions in 30 parallel walking cases, each of which is divided into 10 observing time window of 10 s. The length of sliding window Δt ranges from 1 to 6 s and the interaction orientation threshold is set as an empirical value ranging from 31 to 34. Specifically, the accuracy for interaction detection depends on the thresholds of two acoustic features, i.e., RMS and entropy. We tested their effectiveness in different environments, i.e., an office building that is quiet and a shopping mall with soft or loud music. Results indicated that in a quiet office building, an accuracy of

Fig. 6.6 Group structure at different times

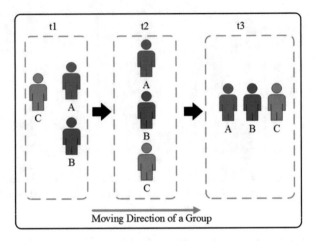

94.4% for interaction detection was obtained while the RMS threshold is 0.009. For a shopping mall with soft music, an accuracy of 90.48% can be obtained while the threshold of spectral entropy is 14.5. Similarly, for a shopping mall with loud music, an accuracy of about 80% is obtained while the threshold of spectral entropy is 14.2. In our experiments, both RMS and spectral entropy of noise are smaller than those of acoustic interactions.

In summary, the proposed approach can be used to recognize the structure of a moving group, i.e., leader-follower relations and left-right relations. Specifically, we first identify the direct leaders in a group. For example, user A is the leader of user C, and user A is in parallel with user B at time t_1, as shown in Fig. 6.6.

Afterwards, orientation changes caused by interactions between any two members can be discovered, which can help recognize the left-right relations. We are able to know that user A is on the left side of user B at time t_1. Finally, the unique group structure is recognized as the final output. With this method, changes in group structures during walking can also be detected. As illustrated in Fig. 6.6, the group structure is side by side at time t_2, and changes to leader-follower at time t_3 because of obstacles in the road.

6.2   Recognition of Group Semantic Interactions[2]

Professionals agree that as much as 50% of meeting time is unproductive and up to 25% is spent discussing irrelevant issues [48]. For example, during a group decision-making meeting, to remain cohesive, participants might not express their opinions;

---

[2]Part of this section is based on a previous work: Z. Yu, X. Zhou and Y. Nakamura, "Extracting Social Semantics from Multimodal Meeting Content," in IEEE Pervasive Computing, vol. 12, no. 2, pp. 68–75, April–June 2013. DOI: https://doi.org/10.1109/MPRV.2012.55

instead they succumb to authority or to the majority opinion—often leading to groupthink [49]. On the other hand, if the floor is open for discussion, fierce debates can occur, undermining group cohesiveness.

Meetings are a human-collaboration process encapsulating a large amount of social and communicative information. The key challenges in creating effective collaborations might be social rather than technical, owing to the dynamics of information exchange [50]. Social factors are thus a fundamental aspect of meeting efficacy. Here, we report on our study of extracting social semantics from meeting content. We developed a three-layer framework and various mining algorithms to gather multimodal meeting content, recognize individual social semantics, and mine the data for group semantics. We also present several potential applications for exploiting the social semantics extracted.

6.2.1 Social Semantics

A few works have been proposed to understand social semantics. Chang et al. [51] utilized several external image/video archives to generate semantic representation in multimedia events. Zhu et al. [52] proposed a novel unsupervised visual hashing approach for semantic image retrieval. Shah et al. [53] presented a multimodal analysis of user-generated data for better semantic understanding of the multimedia content. Wu et al. [54] attempted to understand large-scale video using object and scene semantics. And Heilbron et al. [55] proposed to detect actions in video by embracing semantic priors associated with human activities. The goal of this study was to explore the social semantics of meetings—that is, the human interaction or social behaviors among meeting participants with respect to the current topic, such as proposing an idea, commenting, expressing a positive opinion, or requesting information [56]. We divided the social semantics into two levels: a lower (individual) level and higher (group) level. The lower level semantics refer to the individual interaction occasion (whether an interaction is spontaneous or reactive) and the interaction type, recognized from the multimodal meeting content. The higher level semantics are patterns of group interaction flow and interaction network, mined from the lower level semantics.

Social semantics are particularly important for understanding the social dynamics occurring during a meeting. Is the group converging (coming to the same conclusion) or diverging (presenting many different opinions)? Is one person dominating the meeting and influencing others, or is there balanced participation? Extracting social semantics is useful for different users:

- Meeting participants: Social semantics help make meeting members aware of their own and others' behavior during a discussion (for example, one person proposes or comments a lot while someone else is always critical). This helps members adjust their behavior to increase group satisfaction with the discussion process.

- Meeting organizers: If the organizers know the current meeting status (such as who is being quiet or extroverted or whether the group is coming to an agreement), they can make adjustments to increase meeting efficiency. For example, they might encourage certain participants to speak up or suggest related topics for discussion.
- Meeting sponsors: Meeting sponsors steer the meeting and care about its conclusions because they are usually the ones sponsoring the outcome (the agreed-on plan or activity). Social semantics help sponsors determine whether the meeting was well organized and the conclusion well reasoned. For more information on social dynamics in meetings, see the related sidebar [57].

To help these three user groups, our framework starts by outlining how to collect multimodal content—including video, audio, and motion data (as shown in Fig. 6.7).

6.2.2   Gathering Multimodal Meeting Content

For our study, we mounted cameras in a meeting room to capture the upper-body motions of meeting participants. We also attached a head-mounted microphone to each participant for audio recording, and we used an optical motion capture system for head tracking.

We then extracted social semantics during four meetings—to plan a trip (18 min), prepare for a soccer match (23 min), purchase a PC (26 min), and develop a new job position (10 min). Each meeting had four participants seated around a table—the same four participants attended the first two meetings, and four different participants attended each of the last two meetings (for a total of 12 participants). To obtain the ground truth data, one person (a master's student) annotated the meetings and labeled the interaction occasions and types.

6.2.3   Recognizing the Social Semantics

Here, we describe the recognition approach for the lower level individual semantics. The first step is to detect the basic features—such as head gestures and speech—in the audiovisual and motion data.

6.2.3.1   Interaction Occasion

Once the basic features have been detected, they are used to infer the interaction occasion. There are two types of interaction occasions: spontaneous or reactive. An interaction can be initiated spontaneously by a person (asking a question, for

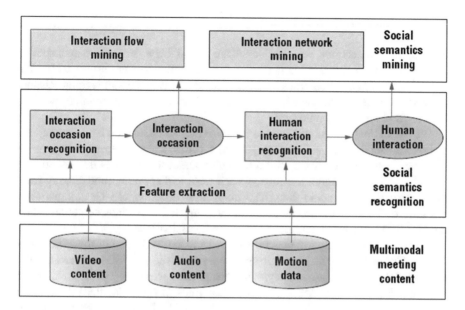

Fig. 6.7 The framework for extracting social semantics from multimodal meeting content

example, is usually spontaneous) or triggered in response to another interaction (acknowledgement, for example, is a reactive interaction).

We adopted an automatic multimodal approach for inferring interaction occasions in multiparty meetings that integrates a variety of features, such as

- Head gesture (nodding, shaking, or normal)
- Attention from others (the number of people looking at the target person during the interaction)
- Attention towards others (the number of people the speaker looks at during the interaction)
- Speech tone (question or nonquestion)
- Speaking time (number of minutes the target person speaks)
- Lexical cue (whether there are any keywords in common with the previous interaction)

The features are extracted from the raw audiovisual and motion data. For example, speech tone and speaking time are automatically determined using the Julius speech recognition engine (http://julius.sourceforge.jp/en). We used a support vector machine (SVM) classifier to classify interaction occasions based on these features.

We captured 1406 interactions from the four meetings, which we used to evaluate the performance of interaction-occasion recognition. We got an overall recognition rate of 0.853 (0.870 for the job selection meeting and approximately 0.84 for the other meetings). The reactive interactions were easier to recognize than the spontaneous ones (with recognition rates of 0.872 versus 0.715).

6.2.3.2   Human Interaction

The human interactions we considered are social behaviors or communicative actions, taken by meeting participants, corresponding to the current topic. We created a set of interaction types based on a standard utterance-unit tagging scheme: propose, comment, acknowledgement, request information, ask opinion, positive opinion, and negative opinion [56].

We used a multimodal method to recognize human interaction based on various audiovisual and high-level features—such as gestures, attention, speech tone, speaking time, interaction occasion, and information about the previous interaction. We extracted the head gesture, attention, speech tone, and speaking time directly from the raw motion and audio data. The interaction occasion can be automatically recognized. To infer the interaction type, we selected four kinds of classification models: SVM, Bayesian net, naive Bayes, and decision tree.

All four inference algorithms correctly recognized over 65% of the 1406 interactions. The statistical significance testing (t-test) showed that SVM outperforms the other models (t was approximately 4, and the p-value was less than 0.01). With the same training set, SVM achieved recognition rates of 0.813 for the PC purchase meeting, 0.740 for the trip planning meeting, 0.736 for the soccer preparation meeting, and 0.804 for the job selection meeting.

6.2.4   Mining Social Semantics

Once the interaction occasion and interaction type have been recognized, the mining algorithms we designed can extract higher level group semantics about the flow and network of interactions.

6.2.4.1   Interaction Flow

An interaction flow is a list of all interactions in a discussion session, with triggering relationships between them. A session is a unit of a meeting that begins with a spontaneous interaction and concludes with an interaction that is not followed by any reactive interactions. Here, spontaneous and reactive interactions can be automatically recognized, as described earlier. A session thus contains at least one (spontaneous) interaction. A meeting discussion comprises a sequence of sessions, in which participants discuss topics continuously.

To form interaction flows in the sessions we also need to specify the interaction to which a reactive interaction is responding. This is manually annotated in our current system, but we plan to explore automatic recognition methods in the future.

We represent the flow of human interaction in a discussion session as a tree. An interaction tree is a rooted, directed, labeled, and acyclic connected graph. Labels are abbreviated names of interactions:

- PRO—propose
- COM—comment
- ACK—acknowledgement
- REQ—request information
- ASK—ask opinion
- POS—positive opinion
- NEG—negative opinion

Figure 6.8 shows one example of interaction trees. The circles represent interactions in the meeting, and the arrows denote an interaction (the start of an arrow) that triggers another interaction (the end of the arrow). In Fig. 6.8, a PRO triggers two interactions, COM and ACK, and then an ACK responds to the COM interaction.

With the interaction flow representation, we designed tree-based mining algorithms to discover frequent patterns of interaction trees as well as frequent subtree patterns [58]. We designed the algorithms to analyze the tree structures and extract interaction flow patterns. To obtain a correct dataset for semantics mining, we tuned the interaction types manually after applying the recognition method. Our data set of the 1406 interactions included 356 sessions—that is, 356 interaction trees.

In frequent interaction tree mining, we set the support threshold at 2% and obtained six patterns. The "support" refers to the ratio of the number of occurrences of a tree to the total number of trees in a dataset. If the support value is larger than a threshold, the corresponding tree is regarded as a pattern. Figure 6.9 presents the patterns, ranked according to their support values (so the support of the tree in Fig. 6.9a is larger than that of Fig. 6.9b, and so on). The support of a tree or subtree T

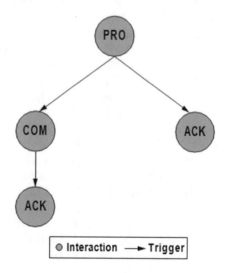

Fig. 6.8   A tree representation of an interaction flow (see the main text for a description of the abbreviations)

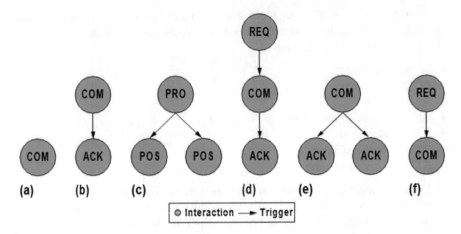

Fig. 6.9 Frequent interaction trees: (**a**) a one-interaction session, in which a person provides information on a topic; (**b**) a person comments on a topic and another person acknowledges the comments; (**c**) a person puts forward a proposal and two other people approve it; (**d**) a person asks a question about a topic, another person answers it, and an acknowledgement follows; (**e**) a person comments on a topic and two acknowledgements are given; and (**f**) a person comments on a topic but there is no acknowledgement

is defined as the value of the number of occurrences of T divided by the total number of trees in a dataset of trees.

From these patterns, we observe that sessions often started with comments, proposals, or questions. Although the most frequent tree is a one-interaction session, most of the frequent trees consist of two or three interactions. Within the six frequent trees, the comment interaction appears five times whereas acknowledgement occurs four times, which means they are common in meeting discussions.

We also examined the patterns extracted from each meeting and found that they overlap. The top three frequent interaction trees of the trip planning meeting, soccer-preparation meeting, and PC purchase meeting are those shown in Fig. 6.9a, c, f; Fig. 6.9a, b, d; and Fig. 6.9a, b, e, respectively. Only one of the top three frequent interaction trees of the job selection meeting is not included in Fig. 6.9 (the other two are those shown in Fig. 6.9e, f).

Setting the support threshold to 4%, we obtain 34 frequent subtrees, of which 7 trees have one node, 12 have two nodes, 10 have three nodes, and 5 have four nodes. Figure 6.10 depicts the top five patterns for each number of nodes ranked by the value of support. We also examined the running time required for the tree-based subtree mining. It took approximately 0.7 s to build the trees and 1.8 s to discover frequent subtrees, which indicates that data mining actually works.

The patterns of one node show the distribution of different interaction types. The comment interaction is the most frequent interaction. Acknowledgement and propose are also very common. The most infrequent interactions are askOpinion and negOpinion, which are not included in the top five. This result indicates that the participants usually did not like to ask someone else's opinion about a proposal or disagree with others.

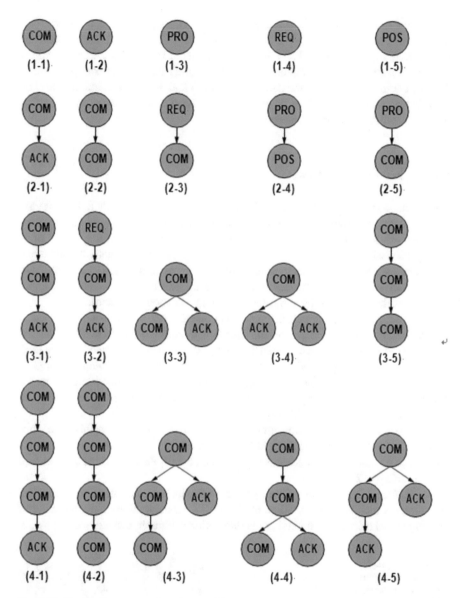

Fig. 6.10 The top five patterns for subtrees with one, two, three, and four nodes (numbers under the pattern indicate the number of nodes and the rank—for example, pattern 3–4 is the fourth rank of three-node patterns)

The two-node patterns are basically combinations of one-node patterns. For example, pattern 2–1 is a combination of 1–1 and 1–2. They also suggest which interaction often follows another one (for example, an acknowledgement often follows a comment).

The patterns of three nodes are combinations of one-node and two-node patterns—for example, pattern 3–2 is composed of 1–2 and 2–3. It can also be constructed with patterns 1–4 and 2–1. The patterns of four nodes are combinations of one-node and three-node patterns or two two-node patterns. For example, pattern 4–1 is a combination of 2–1 and 2–2 as well as 1–1 and 3–1. These patterns could be regarded as part of the grammar of a meeting discussion. Discussions following these interaction flows might be more efficient.

6.2.4.2   Interaction Network

In a discussion session, meeting participants interact with each other. The interactions construct a network—that is, an interaction network. Investigating the interaction network is useful to extract the semantic knowledge for learning social rules during discussions, such as how many people from a group are usually involved in a discussion session, how many interactions occur in a discussion session, and whether any subgroup interactions occur (for example, do two people often interact with each other, or is one person often responding to another person).

We used a graph to represent an interaction network in a session. Interaction graphs are directed (in that one person responds to another) and labeled and can be unconnected (when there are isolated nodes). The vertices represent the meeting participants. A distinguished vertex is the session initiator, and edges represent interaction responses. Edges are labeled with numbers denoting how many times a person responds to another person during the session. An isolated node means that the person did not interact with anyone.

Two examples of interaction graphs are shown in Fig. 6.11. In this case, four people participated in the meeting. In Fig. 6.11a, P1 initiates the session, P2 and P3 respond to P1, and then P1 responds to P2. In Fig. 6.11b, P2 initiates the session, and P1 responds to P2 two times.

To explore subgroups in a discussion, we introduce the notion of an interaction subgraph—which is a connected graph. Unlike in the interaction graph, the initiator and edge label are not considered in the interaction subgraph; instead, we focus only

Fig. 6.11 A graph representation of the interaction network: (**a**) P1 initiates the session, P2 and P3 respond to P1, and P1 responds to P2. (**b**) P2 is the session initiator, and P1 responds to P2 two times

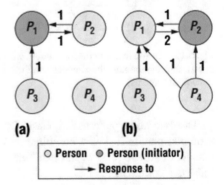

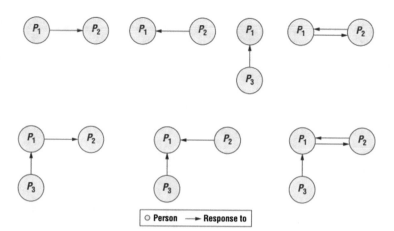

Fig. 6.12 Interaction subgraph examples

on how many people interact and who interacts with whom, not on who is the initiator and how many times a person responds to another person. Figure 6.12 shows all subgraphs of the interaction graph depicted in Fig. 6.11a.

We designed graph mining algorithms to discover frequent patterns of interaction graphs and subgraphs. To illustrate interaction network mining, we used the trip planning and soccer preparation meetings (totaling 41 min). Both meetings had the same four participants. The graph dataset contains 251 interaction graphs.

In frequent interaction graph mining, we set the support threshold at 3% and obtained six patterns (see Fig. 6.13). Figure 6.13a shows the most frequent graph, which denotes P3 launching the session and P4 responding to P3. The patterns shown in Fig. 6.13b, c represent one-interaction sessions. The pattern of Fig. 6.13d represents P4 initiating the session and P3 responding to P4, and patterns of Fig. 6.13e, f represent three-interaction sessions. From these patterns, we can conclude that sessions were often launched by P3 or P4. They were active in the meetings and might be central in the group. P1 was the least active.

In subgraph mining, with the same support threshold setting (3%), we got 53 frequent subgraphs, in which 12 graphs have one edge, 30 have two edges, and 11 have three edges. The top five patterns for each number of edges, ranked by the support value, are presented in Fig. 6.14.

The one-edge patterns represent the most primitive interaction patterns. The patterns of two edges are basically combinations of one-edge patterns. For example, pattern 2-1 is a combination of 1-1 and 1-2. Pattern 2-2 is obtained from 1-3 and 1-4. The patterns of three edges are combinations of one-edge and two-edge patterns. For example, pattern 3-1 is a combination of 1-4 and 2-3. It can also be obtained from 1-3 and 2-5.

From the patterns discovered with subgraphs, we can conclude that P3 and P4 will likely form a clique in a discussion—that is, they often respond to each other. In addition to this clique, P2 often connects with P3 or P4 and forms less frequent subgroups. These extracted patterns can be used to interpret human interaction in small group discussions.

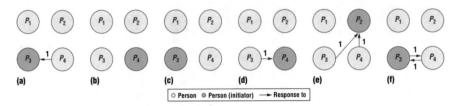

Fig. 6.13 Frequent interaction network patterns: (**a**) the most frequent graph, which denotes P3 launching the session and P4 responding to P3. Another pattern is that of one-interaction sessions, with (**b**) P4 or (**c**) P3 as the initiator. (**d**) P4 initiates the session and P3 responds to P4. There are also three-interaction sessions with (**e**) P2 or (**f**) P3 as the initiator

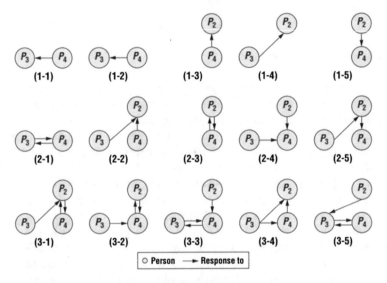

Fig. 6.14 Top five patterns for interaction subgraphs with one, two, and three edges (numbers under the pattern indicate the number of edges and the rank—for example, pattern 3-2 is the second rank of three-edge patterns)

6.2.5   Applications

Several potential applications could be developed based on the social semantics extracted.

Meeting Browser: The extracted social semantics can be incorporated into a meeting browser—that is, a visualization tool for browsing information of a recorded meeting—to help people quickly understand the social interactions occurring in the meeting. This could help improve meeting organization and participation. For example, with the extracted social semantics, the organizer could grasp the current meeting status and make necessary adjustments to improve meeting efficiency. Social psychologists suggest that a group can improve its interaction and consequently its productivity by understanding its emotional and social interactions

[59]. In addition, balanced participation is essential for proper problem-solving. Browsing the social interaction, behavior, and network could help make members aware of their own and others' behavior in a discussion so changes could be made to increase group satisfaction with the discussion process. An individual, a group, or even an entire organization could benefit from the participants' awareness of their own behavior during meetings [60].

Smarter Indexing: Indexing is useful for efficiently organizing, managing, and storing meeting data. Meaningful indices act as semantic pointers into the meeting record. The extracted social semantics could be adopted to develop a smarter indexing tool for querying different social aspects of the meetings. For example, using interaction recognition, you could index the person who proposed the most ideas or made the most critical comments, or you could index the most common or controversial topics. Furthermore, the system could query higher level mining results to identify typical interaction flows and subgroup interactions.

Cognitive Analysis: Another potential application is to use the extracted social semantics for cognitive analysis. The extracted higher level semantics, such as interaction flow patterns, are extremely useful for interpreting human interaction in meetings. Cognitive science researchers could use the patterns as domain knowledge for further analysis of human interaction. Moreover, the discovered patterns could be used to evaluate whether a meeting discussion was efficient.

Although our interaction recognition was acceptable, we are still in the early stages of development. In the future, we plan to integrate more contexts into the recognition process to improve the recognition accuracy. We also plan to incorporate more meeting content, in terms of both volume and type. The meetings used in this study were all task oriented, but we would also like to analyze other meetings, such as panel discussions, debates, or interviews, to identify differences in the frequent interaction patterns for different meeting styles.

6.3   Recognition of Group Interaction Patterns[3]

6.3.1   Introduction

Meetings are important events in our daily lives for purposes of problem-solving, information exchange, and knowledge sharing and creation. Thanks to pervasive computing technologies, meetings tend to be smart. A smart meeting system can assist people in a variety of tasks, such as scheduling meetings, taking notes, sharing files, and browsing minutes after a meeting [57, 61–65].

---

[3]Part of this section is based on a previous work: Z. Yu, Z. Yu, H. Aoyama, M. Ozeki and Y. Nakamura, "Capture, recognition, and visualization of human semantic interactions in meetings," 2010 IEEE International Conference on Pervasive Computing and Communications (PerCom), Mannheim, 2010, pp. 107–115. DOI: https://doi.org/10.1109/PERCOM.2010.5466987

Current smart meeting systems mainly support meeting preparation (before meeting), information exchange (in meeting), and content review (post-meeting) [66], whereas little research has been conducted on social aspects. Meetings encapsulate a large amount of social and communication information. Group social dynamics is particularly important for understanding how a conclusion was reached, e.g., whether all members agreed on the outcome, who did not give his/her opinion, and who spoke a little or a lot. They are further useful for determining whether the meeting was well organized and the conclusion well reasoned. An important characteristic of group social dynamics is human interactions.

Unlike physical interactions such as turn-taking and addressing (who speaks to whom), the human interactions considered here are defined as social behaviors among meeting participants with respect to the current topic, such as proposing an idea, giving comments, expressing a positive opinion, and requesting information. When incorporated with semantics (i.e., user intention or attitude towards a topic), interactions are more meaningful for evaluating conclusion drawing and meeting organization. Understanding how people are interacting can also be used to support a variety of pervasive systems [67]. For example, previous meeting capture and access systems could use this technology as a smarter indexing tool to access different parts of the meetings.

In this section, we focus on manipulation of this type of human semantic interactions in meetings, specifically capturing, recognizing, and visualizing them. A smart meeting system was built for this purpose. We adopt a collaborative approach for capturing interactions by employing multiple sensors, such as video cameras, microphones, and motion sensors. A multimodal method is proposed for interaction recognition based on a variety of contexts, such as head gestures, attention from others, speech tone, speaking time, interaction occasion, and information about the previous interaction. A support vector machine (SVM) [68] classifier is used to classify human interactions based on these features.

6.3.2 Related Work

A number of smart meeting systems have been developed in the past using pervasive computing technologies. The Conference Assistant [61] combines context awareness and wearable computing technologies for enhancing attendee interactions with the environment and other attendees. EasyMeeting [63] aims at providing services to meeting participants based on their situational requirements, such as preparing the data projector and setting the lighting and temperature in the room. The FXPAL conference room [65] monitors a meeting using a variety of devices and provides tools for accessing and browsing captured meetings. TeamSpace [66] offers supports of meeting preparation, capture, and access for distributed workgroups. Ahmed et al. [69] built a smart meeting room that detects the beginning and end of a meeting and supports ephemeral group communication. The Intelligent Meeting Room [70] recognizes activities in a meeting room, such as a person locating in front of the whiteboard, a lead presenter speaking, and other participants speaking. SMeet [71] enables multiparty meeting by integrating user-friendly pointing/tracking interactions, high-resolution tiled displays, hybrid multicast-based networking, and high-quality media services. Araki et al. [72, 73] studied in speech enhancement for online meeting recognition system. They proposed models to deal with speaker overlaps and noise signals, which could achieve better performance of speech recognition in meetings. The shared goal of these works was making an intelligent meeting room that provides adaptive services and relevant information using ubiquitous sensors and context awareness techniques. This differs from our study, which focuses on capturing human interactions and understanding their semantic meaning.

Numerous studies have focused specifically on physical interaction recognition and visualization in meetings. Yun et al. [74] proposed to detect interaction between two persons using body-pose features and multiple-instance learning. Ji et al. [75] presented an interactive body part model and calculated the spatial-temporal joint features for eight interactive limb pairs in a short frame set for motion description. Zhu et al. [76] applied neural network framework with co-occurrence feature learning for skeleton-based action recognition. Stiefelhagen et al. [77] proposed an approach for estimating who was talking to whom, based on tracked head poses of the participants. AMI project [78] deals with interaction issues, including turn-taking, gaze behavior, influence, and talkativeness. The Meeting Mediator at MIT [79] detects overlapping speaking time and interactivity level in a meeting by using Sociometric badges and then visualizes the information on mobile phones. Sumi et al. [80] analyzed user interactions (e.g., gazing at an object, joint attention, and conversation) during poster presentation in an exhibition room. DiMicco et al. [81] presented visualization systems for reviewing a group's interaction dynamics, e.g., speaking time, gaze behavior, turn-taking patterns, and overlapped speech in meetings. Otsuka et al. [82] used gaze, head gestures, and utterances in determining interactions regarding who responds to whom in multiparty face-to-face conversations. Although several other systems, e.g., [83–85], seem to model and analyze interactions, they deal with very-high-level group actions, such as presentation, general discussion, and note-taking. In general, the abovementioned systems mainly focus on the analysis of physical interactions between participants without any relation to the topics; in other words, they do not include semantic meanings in the analysis. Therefore, they cannot determine clearly a participant's attitude or role in a discussion.

A few systems attempted to analyze semantic information in meeting interactions. Hillard et al. [86] proposed a classifier for the recognition of an individual's agreement or disagreement utterances using lexical and prosodic cues. The Discussion Ontology [87] was proposed for obtaining knowledge such as a statement's intention and the discussion flow in meetings. Both systems are based on speech transcription that extracts features such as heuristic word types and counts. Garg et al. [88] proposed an approach to recognize participant roles in meetings. Laurent et al. [89] also proposed a tree-based machine learning algorithm to identify speaker roles. The speech act theory [90] determines implicit actions (e.g., greeting, apologizing, requesting, promising) behind human speech based on utterances. Our system differs in several aspects. First, besides detecting positive or negative opinions [86], we analyze more types of human semantic interactions in topic discussion such as proposing an idea, requesting information, and commenting on a topic. Second, our system adopts a multimodal approach for interaction recognition by considering a variety of contexts (e.g., head gestures, face orientation, speech tone), which provides robustness and reliability. Third, we present a visualization interface for browsing interactions, as well as relationships between them.

Another area of related work is about meeting browser, which has its root in Waibel et aI.'s work on summarizing and navigating meeting content [91]. Afterwards several other similar systems were developed. Rough'n'Ready [92] provides audio data browsing and multivalued queries, e.g., specifying keywords as topics, names, or speakers. Ferret [93] offers interactive browsing and playback of many kinds of meeting data including media, transcripts, and speaker segmentations. Geyer et al. [94] discussed various ways of indexing meeting records and built a viewer for accessing the content. Jaimes et al. [95] proposed a meeting browser enhanced by human memory-based meeting video retrieval. Junuzovic et al. [96] presented a 3D interface for viewing speaker-related information in recorded meetings. Besides meeting content that existing systems are mainly dedicated to, our visualization system provides a multimodal and interactive way for browsing human interactions.

6.3.3 Human Semantic Interaction

Meetings usually encapsulate a large amount of communicative statements and semantic relationships between them that form an interaction network. To enrich knowledge about a meeting with social group dynamics, we need to analyze not only physical interactions such as turn-taking and addressing (who speaks to whom) between participants, but also semantic meanings behind the physical actions.

Human semantic interactions are defined as social behaviors among meeting participants with respect to the current topic. Various interactions imply different user roles, attitudes, and intentions about a topic during a discussion. With semantics incorporated, interactions are more meaningful in understanding conclusion drawing and meeting organization.

The definition of interaction types naturally varies according to usage. For generalizability, we create a set of interaction types based on a standard utterance-unit tagging scheme [97]. It includes the following seven categories of human semantic interactions: propose, comment, acknowledgement, requestInfo, askOpinion, posOpinion, and negOpinion. The detailed meanings are as follows: propose-a user proposes an idea with respect to a topic; comment-a user comments on a proposal, or answers a question; acknowledgement-a user confirms someone else's comment or explanation, e.g., "yeah," "uh huh," and "OK"; requestInfo-a user requests unknown information about a topic; askOpinion-a user asks someone else's opinion about a proposal; posOpinion-a user expresses a positive opinion, i.e., supports a proposal; and negOpinion-a user expresses a negative opinion, i.e., disagrees with a proposal.

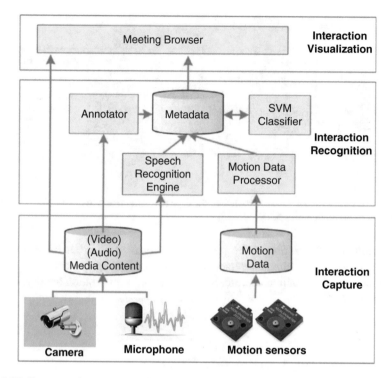

Fig. 6.15 Smart meeting system architecture

6.3.4   System Architecture

We developed a prototype smart meeting system for capture, recognition, and visualization of human semantic interactions during a discussion. Figure 6.15 illustrates the system architecture, which consists of the following three layers: interaction capture, interaction recognition, and interaction visualization.

In this design, *interaction capture* is the physical layer that deals with capture environment, devices, and methods. Video cameras, microphones, and motion sensors are used for recording meeting content and tracking participants' head movement.

Interaction recognition serves as the structural layer, which analyzes contents of the generated audiovisual data and motion data. The speech recognition engine extracts speech features (tone and speaking time) from audio data. The motion data processor is responsible for analyzing sensing data derived from motion sensors, and measuring people's head gestures (e.g., nodding) and face orientation. The annotator is used for manually labeling features from audiovisual data. Based on low-level features, the SVM classifier recognizes human interactions. The interaction recognition layer makes the meeting content meaningful and provides support for the upper layer.

Fig. 6.16 Capture system: (**a**) an overview of the capture environment; (**b**) a participant wearing a microphone and motion sensors

Interaction visualization is the presentation layer, offering an interactive user interface for browsing human interactions as well as meeting content.

The details of interaction capture, recognition, and visualization are described in the following sections.

6.3.5   Collaborative Interaction Capture

We use multiple devices for capturing human interactions. Figure 6.16 is a snapshot of our capturing system setup. An overview of the capture environment is shown in Fig. 6.16a; Fig. 6.16b shows a participant wearing a microphone and motion sensors.

6.3.5.1   Video Capture

Six video cameras (SONY EVI-D30) are deployed in our capture system. Four capture the upper-body motions of four participants (breast shot), one camera records presentation slides (screenshot), and the other captures an overview of the meeting including all participants (overview shot).

The breast-shot camera is controlled automatically based on face recognition of its assigned participant. The face region is extracted from the image of each breast-shot camera using face recognition software (provided by Toshiba Ltd.). If the center of the face does not coincide with the center of the image, the camera is adjusted until the face appears at image center. If the face is already at the center, the camera is zoomed in or out so that the appropriate size of the face in the image is obtained.

The video signal from each camera is input to an encoder board and stored in the MPEG2/PS format with a frame size of 720×480 pixels, frame rate of 29.97 fps, and bit rate of 2 Mbps.

6.3.5.2 Audio Capture

To capture audio data, a head-mounted microphone (SHURE WH30XLR) is attached to each participant. A fixed microphone records global meeting sound. The audio signal (MPEGI-Layer II 48,000 Hz [CBR Stereo] 384 kbps) is also imported into the encoder board in sync with the video signal. The computers attached with the encoder boards are synchronized with each other by network time protocol (NTP).

6.3.5.3 Head Tracking

We use an optical motion capture system (PhaseSpace IMPULSE) [98] for head tracking. It mainly consists of three parts: light-emitting diode (LED) module, camera, and server. The tracking system uses multiple charge-coupled device (CCD) cameras for three-dimensional measurement of LED tags (sensors) and obtains their exact position. We placed six LED tags around each person's head. The LED tags are scanned 30 times per second, and position data can be obtained in real time with less than IO-ms latency. Through the three-dimensional position data, head gestures (e.g., nodding) and face orientation can be detected.

6.3.6 Multimodal Interaction Recognition

6.3.6.1 Context Extraction

The contexts considered in our interaction recognition include head gestures, attention from others, speech tone, speaking time, interaction occasion, and type of the previous interaction. These features are extracted from the raw audiovisual data and motion data.

Head Gesture Recognition: Head gestures (e.g., nodding and shaking of the head) are very common, and are often used in detecting human response (acknowledgement, agreement, or disagreement). We determine nodding through the vertical component of the face vector calculated from the position data. The nodding recognition method is schematically shown in Fig. 6.17a.

We first determine the maximum and minimum value of the vertical component of the face vector in a time window (here we set 1 s). Next, we calculate 81 (the difference between the maximum and minimum values) and 82 (the difference between the current value and the minimum). Then the nodding score is calculated using Eq. (6.8):

$$\text{Score} = \frac{\theta_1}{11.5} \times \frac{\theta_2}{11.5},$$

$$(6.8)$$

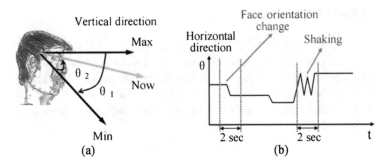

Fig. 6.17 Head gesture recognition method: (**a**) nodding; (**b**) shaking

where 11.5 is empirically set as the normalization constant. If the calculated score is above the preset threshold (e.g., 0.80), we consider that the head gesture is nodding.

Headshaking is detected through the horizontal component of the face vector. The method is illustrated in Fig. 6.17b. We first calculate a projection vector of the face vector on the horizontal plane. To distinguish headshaking from normal face orientation change, we count the switchback points of the projection vector's movement in a time window (e.g., 2 s). If the number of the switchback points is larger than the preset threshold (e.g., 2), we consider that it is a shaking action.

Detection of Attention from Others: Attention from others is an important determinant of human interaction. For example, when a user is proposing an idea, he/she is usually being looked at by most of the participants. The number of people looking at the target user during the interaction can be determined by their face orientation. We measure the angles between the reference vectors (from the target person's head to the other persons' heads) and the target user's real face vector (calculated from the position data). Face orientation is determined as the one whose vector makes the smallest angle.

Speech Tone and Speaking Time Extraction: Speech tone indicates whether an utterance is a question or a non-question statement. Speaking time is another important indicator in detecting the type of the interaction. When a user puts forward a proposal, it usually takes relatively longer time, but it takes a short time to acknowledge or ask a question. Speech tone and speaking time are automatically determined using the Julius speech recognition engine [99]. It segments the input sound data into speech durations by detecting silence interval longer than 0.3 s. We classify segments as questions or non-questions using the pitch pattern of the speech based on hidden Markov models [100] trained with each person's speech data. The speaking time is derived from the duration of a segment.

Identification of Interaction Occasion: An interaction is either spontaneous or reactive. In the former case, the interaction is initiated spontaneously by a person (e.g., proposing an idea or asking a question). The latter denotes an interaction triggered in response to another interaction. Discussion tags [87] can be used to explicitly indicate the interaction occasion. We manually label this feature in our current system using Anvil [101], which is a widely used video annotation tool with user-defined hierarchical layers and attributes.

Detection of Type of Previous Interaction: The type of the previous interaction also plays an important role in detecting the current interaction. It is intuitive that certain patterns or flows frequently occur in the course of a discussion in a meeting. For instance, propose and requestInfo are usually followed by a comment. This feature can be obtained from the recognition result of its previous interaction.

6.3.6.2   Interaction Recognition Based on Support Vector Machines

With the basic context aggregated, we adopt the SVM classifier for interaction recognition. Given the features of an instance, the SVM classifier sorts it into one of the seven classes of interactions. SVM has been proven to be powerful in classification problems, and often achieves higher accuracy than other pattern recognition techniques. It is well known for its strong ability to separate hyperplanes of two or more dimensions.

Our system uses the LIBSVM library [102] for classifier implementation. The default kernel type is the radial basis function (RBF): $K(x_i, x_j) = \exp{(-\gamma\|x_i - x_j\|^2)}$, $\gamma > 0$. Two parameters ae important while using LIBSVM: C and γ. Good (C, γ) enables the classifier to accurately predict unknown data (i.e., testing data). In our system, we adopt a "grid search" to find the best parameter C and γ using cross-validation. Basically, pairs of (C, γ) are tried and the one with the best cross-validation accuracy is picked.

The meeting content is first segmented into a sequence of interactions. Sample interactions are selected and fed to the SVM as training data, while others are used as a testing set.

6.3.7   Performance Evaluation

We evaluate the performance of the developed system from two aspects, i.e., the efficacy of context extraction and the efficacy of interaction recognition. The used data includes two meetings, one soccer preparation meeting (23 min, talking about the players and their role and position in a coming match) and one trip planning meeting (18 min, discussing about time, place, activities, and transportation of a summer trip).

6.3.7.1   Evaluation of Context Extraction

The evaluation of context extraction mainly includes detection of head gestures (nodding and shaking), speech tone (questions), and attention from others (face orientation). The extraction accuracies of these three types of contexts are 76.4%, 80.0%, and 72.2%, respectively, as reported in Table 6.2. The recognition rate of question tone is a little low. The reason might be that the Julius speech recognition

Table 6.2 Context extraction results

Context	Recognition rate (%)
Head nodding	76.4
Headshaking	80.0
Question tone	65.0
Face orientation	72.2

Table 6.3 Results of different interaction recognition

Interaction type	Recognition rate (%)
Propose	22.2
Comment	76.3
Acknowledgment	86.7
requestInfo	83.3
askOpinion	0.0
posOpinion	80.0
negOpinion	0.0
Total	74.3

engine omitted numerous short sentences, while questions usually contain short sentences. It also verified that speech recognition in meeting is challenging due to highly conversational and noisy nature of meetings, and lack of domain-specific training data [103].

6.3.7.2   Evaluation of Interaction Recognition

We further evaluated interaction detection by measuring the recognition rate. Based on the video and audio data, we manually labeled 518 interactions in the soccer meeting. We chose 370 of them as a training set, and the other 148 were used for testing. As summarized in Table 6.3, an average recognition rate of 74.3% was obtained. We can find that the proposed algorithm performs well to recognize some of the interactions, including comment, acknowledgment, requestInfo, and posOpinion. In other words, these four types of interactions can be easily detected based on the used features. On the other hand, the other three types of interactions (propose, askOpinion, and negOpinion) are difficult to recognize. Specifically, some proposed interactions were wrongly classified into the category of comment and all askOpinion interactions were recognized as requestInfo.

We then used the training data of soccer meeting to test the trip meeting and achieved an accuracy of 72.6%. The performance was almost equal to the result of training and testing with data of the same meeting (74.3%). This indicates that a model trained based on one meeting can be used to test other meetings. For more details about the evaluation results, please refer to the article [104].

6.4 Group Activity Organization and Suggestion with Mobile Crowd Sensing[4]

6.4.1 Introduction

Many technologies facilitate group interaction [105]. For example, group management tools [106–110] can help analyze historical interaction data of online communication (e.g., email and Facebook) and off-line collocated social events [111–115]. However, group activities involve more complex processes such as group formation and event publicity in addition to intragroup interaction during the events.

This is a challenging problem. First of all, different types of group activities may vary in goals, needs, constraints, flows, organizations, and interaction patterns. For instance, some events are closed or private (e.g., a party). In other cases, activities are open to the public. In addition, each organizational stage of an activity may require different technical support. For example, one challenge of activity preparation is to locate and invite potential attendees. In contrast, the core issue of running an activity is recognizing and monitoring ongoing events. Group would benefit from a conceptual model for automatic processing. Second, there is a lack of technical infrastructure for group activity logging and mining. For online communities, social web portals can capture virtual interaction data for further use, such as social tie detection. Data from real-world group activities are harder to obtain, requiring specialized models, methods, and mechanisms. It is more difficult to extract information and infer knowledge from physical group activities, since the data tend to be noisy and incomplete. Third, activity organization in real-world settings is often influenced by various social and physical contextual factors, such as user location, activity venue/time, and existing participants.

Mobile crowdsensing (MCS) [116] leverages crowd-contributed data collected via smartphone sensors in the physical space as well as mobile social networks (SNs) in the cyber space. MCS has been employed in numerous application areas, yet its use in group activity organization is under investigation. Our work aims to exploit the cross-space sensing nature of MCS to support the life cycle of real-world group activities.

In this chapter, we present MobiGroup, a group-aware system that provides assistance throughout various group activity organizational stages. It exploits smartphone sensing to capture online/off-line social interactions and empowers group formation and management. We extend [117] by (1) addressing the activity life cycle in real-world settings; (2) characterizing the complexity and diversity of social activity organizational processes in a formal concept model; and (3) providing

---

[4]Part of this section is based on a previous work: B. Guo, Z. Yu, L. Chen, X. Zhou and X. Ma, "MobiGroup: Enabling Lifecycle Support to Social Activity Organization and Suggestion With Mobile Crowd Sensing," in IEEE Transactions on Human-Machine Systems, vol. 46, no. 3, pp. 390–402, June 2016. DOI: https://doi.org/10.1109/THMS.2015.2503290

intelligent facilitations for social activity preparation. Specifically, our contributions include the following.

- A generic and multiviewed group activity model: The activity model classifies the life cycle of group activities into four stages. Based on the model, we develop a framework that can adapt supports to the characteristics and organizational stages of a group activity.
- Context-aware approaches to group activity preparation: For activities that are open to the public, we propose a heuristic rule-based strategy to disseminate information of an activity according to its popularity and group preferences [118]. For private activities that often consist of a similar set of participants, we use a social graph model to characterize the closed activity participation network and develop a context-based group computing method for highly relevant group recommendation.
- Cross-community approach to ongoing activity suggestion: To encourage participation of ongoing events, we propose a mechanism for recommending activities to potentially interested users by extracting static/dynamic interaction features of both online and off-line communities [119].

6.4.2   Related Work

Groupware refers to software that can help people achieve common goals in collaborative work [105]. Examples include email and group editing/conferencing tools [120, 121]. FlierMeet [122] is a crowdsensing system that encourages group of users to achieve goals like information tagging and sharing. FooDNet [123] proposed a crowdsourcing framework for food delivery across a group of users in a city. CrowdWatch [124] presented a novel system that leverages a group of users in a mobile crowdsensing manner to detect temporary obstacles and then make effective alerts for distracted walkers. ContactMap [110] provided an editable group visualization tool to depict personal contacts and groups. SocialFlow [106] displayed social groups mined from email data. Researchers from Google proposed a method that can suggest a recipient group upon email composition [107]. These systems extracted social groups from online interactions. They did not address group activity organization in real-world scenarios.

Mobile group activity augmentation refers to the techniques that aim to augment group activity organization using mobile devices. Many studies focus on group activity sensing and sharing. For instance, CenceMe exploited mobile phones to infer people's presence (e.g., dancing at a party with friends) and then shared this information over SN media [111]. Movi collaboratively took photos and shared the social activities through collocated mobile phones [112]. Pinkerton et al. [125] studied the reasons of why participants would like to share real-world activities on social network platforms. Our work adopts an audio-based smartphone sensing method to recognize social activities. Rather than simply sharing the ongoing activity information with friends online, we take an opportunistic approach to selectively send the message to friends that meet certain criteria (e.g., distance, static intimacy (SI)/dynamic intimacy (DI)). The participant suggestion for real-world activity organizers is another way to augment group activities. Flocks [113] was a system that supports dynamic group creation on the basis of user profiles and physical proximity. SOCKER allowed the building of ad hoc groups based on opportunistic data dissemination [114]. These systems mainly grouped people already located nearby and did not recruit like-minded contacts who are not yet gathered but could be. Furthermore, no existing studies have given a systematic investigation of the generic process of group activity organization. MobiGroup, to the best of our knowledge, provides the first concept model of this field.

People are involved in multiple communities, either online or off-line. Different communities are implicitly interlinked [114, 115]. For example, colocation in the real world is echoed in connectivity online [126]. We view online and off-line communities as complementary networks and leverage "cross-community sensing and mining" techniques [119] to support group activity organization. Several studies explore the integration of features from hybrid communities. For example, Tang et al. [127] transferred and integrated knowledge from different SNs for social tie prediction. Our previous work [128] combined pervasive sensing and Web intelligence techniques for social contact management. MobiGroup facilitates group activity organization by considering both online (e.g., friendship, comments) and off-line features (e.g., location, ambient sounds).

6.4.3 Group Activity Modeling

6.4.3.1 Group Activity Characterization

Group activities can be held in face-to-face or in online manners. In our study, group activities are referred to traditional meeting-based activities, i.e., a crowd of people that gather together at a certain time and place for a specific purpose. We explore the key concepts of group activities in the following.

Group Activity Type: Based on the organization manner, we can divide group activities into planned and unplanned (or opportunistic) activities [129] (see Fig. 6.18). Planned activities usually have explicit activity initiators and require participant invitation or activity advertising, such as organizing a workshop/party. Unplanned activities are usually held in an opportunistic manner and do not have a clear preparation process, such as meeting someone in a store. Based on the openness of an activity to its participants, we can categorize group activities into public/open activities and private/closed ones [130, 131]. In general, public activities are widely announced and do not limit participation. The participants of private activities are relatively static (or fixed) and are not recruited through open calls but by personal invitations. For instance, Bob, a university student, often has his lunch

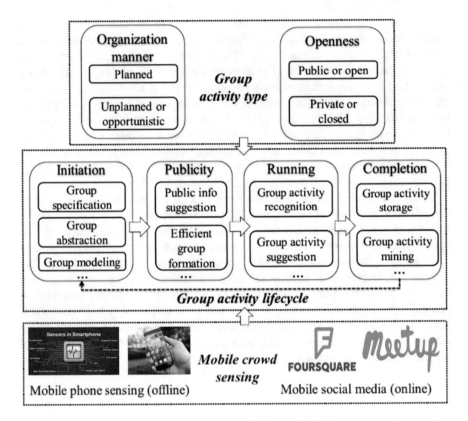

Fig. 6.18 Conceptual graph of group activity organization

with a similar subset of his contacts, e.g., with B, C, and D on one day and C and D on another day.

Activity Life Cycle: Organizing a group activity can be characterized as a four-stage life cycle: activity initiation, publicity, running, and completion. Different technical support should be given to each stage.

1. Initiation and publicity: These two stages are about activity preparation. For public activities, activity advertising and recommendation are critical. For private activities, since the participants are relatively static and fixed, it is desirable to suggest the right members to the private activity initiator.
2. Running: Unplanned activities usually do not have a preparation process and the running stage is crucial in its life cycle. Within this stage, the activity participants can share their status online and the system can suggest relevant people to join them. Depending on the situation, participants may prefer to involve different cohorts of contacts based on distinct intimacy metrics, such as close friends, or the ones interacting frequently recently.

3. Completion: After a group activity completes, its information should be kept in the backend server and used for individual/group pattern learning. The learned information will be helpful to feed the prior three stages.

Activity Tags: To facilitate activity information sharing and advertising, each activity instance will be associated with tags. We use broadly two types of tags: category and semantic tags.

1. Category tags: Category tags are defined based on the nature and task of activities such as meeting, concert, or party.
2. Semantic tags: In addition to category tags, in information systems, we often characterize items by their semantics [132, 133]. We define the following two semantic tags:

   Hot: An activity instance's hotness can be measured by the number of people who interact with it (e.g., pressing the "like" button, reposting messages), and this number usually reaches a peak value within a short period of time.

   Social: People from existing groups usually show high similarity, which is often defined as group preference [118, 134]. This inspires us to characterize an activity instance at the social structure level. For instance, we can recommend a "social" activity to an existing online group if a portion of members from this group likes it. Groups can be extracted using community detection methods [135].

   "Hot" is a global feature, and it indicates the interaction dynamics of the whole community (with loose connections). In contrast, "social" is a local feature, representing the preferences of a group of highly connected people.

6.4.3.2 Group Formation

Another important concept pertaining to group activity organization is the formation of groups. We define three types of groups.

Activity participation groups: Activity participation groups are formed by people who participate in activities in the real world. For public activities, dynamic groups are usually formed through open calls. For private activities, static groups that usually consist of similar sets of social contacts are often formed. The participants of each activity instance form a raw group, and logical groups can be distilled when studying implicit grouping rules and merging similar raw groups based on historical data. For instance, as presented in the above section, the lunch group B, C, and D and C and D can be merged under certain conditions.

Third-party groups: We can use the social relationship and interaction from other communication mediums. As in ref. 119, an understanding and prediction of group activities can be attained by mining the data from heterogeneous sources from cyber, physical, and social spaces. We thus employ third-party groups that can be detected from online cyber portals, e.g., location-based SNs (LBSNs), to assist real-world activity organization. The analysis of third-party groups can be conducted at two levels of abstraction: structure level and interaction level (e.g., comments, likes, retweets). As presented in ref. 131, the SN structure mirrors relatively "static" user connections. It can be analyzed at either the global scale (e.g., the whole community [135]) or the local scale (e.g., ego-network analysis [136]). Interaction-based analysis often reveals dynamic connecting features (e.g., the interaction frequency between two linked users changes over time).

Context-based groups: Context is an important factor for group formation, in particular, in the case of ongoing activities. A widely acknowledged definition of context is "any information that can be used to characterize the situation of an entity [137]." In this study, the entity is group activity, and all information pertaining to activity such as user location, activity time and venue, social intimacy, user preferences, and behavior similarity is a useful context that can impact the formation of activity groups.

Sometimes, we need to combine the features of different group types to achieve hybrid grouping, e.g., the formation of groups that meet both spatiotemporal contexts and high interaction frequency.

6.4.4 MobiGroup Architecture

Based on the study of the group activity model, we have developed MobiGroup to address (1) how to support intelligent advertising of public activities; (2) how to facilitate the suggestion of highly relevant groups to prepare private activities; and (3) how to recommend ongoing activities according to social and physical contexts. The layered architecture of MobiGroup is in Fig. 6.19.

The first layer is the MCS layer, which consists of smartphones enhanced by various sensors (e.g., Bluetooth, Wi-Fi, GPS, accelerometers, and microphone) and third-party SN services.

The second layer is the data collection and storage layer that contains two modules: heterogeneous crowdsourced data collection and activity registration center (ARC). The former provides MobiGroup with gateways to collect the needed data from either smartphone sensors or third-party services, while the latter transforms raw data to social activity logs and inserts them into the ARC repository. The activity logs will be used for historical data-based group extraction and suggestion.

To facilitate group activity organization, we have developed the activity registration model (ARM) for ARC. As shown in Fig. 6.20, in ARM, each activity instance includes an initiator, the initiation place (I-Loc) and time (I_Time), the activity venue (Venue) and time period (A_Time), and a number of activity members or participants (MemL). We use an example to illustrate this model: Bob is in the lab and wants to invite some friends to have dinner together at the Golf restaurant. Here, Bob is the activity initiator (he initiates the activity in his lab), and the activity venue is the Golf restaurant. For unplanned activities that often lack the initiation process, the ARM works as follows: The initiator is by default the current activity sharer and

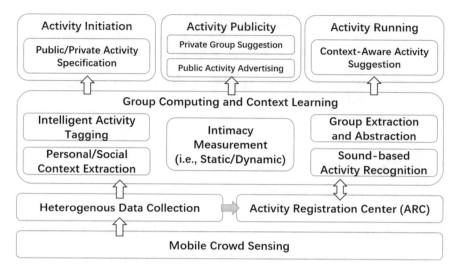

Fig. 6.19 MobiGroup system architecture

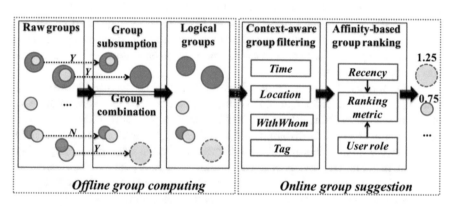

Fig. 6.20 Approach for private activity preparation

the members are those who participate in the activity after suggestion; the initiation place is the same as the activity venue.

The third layer is the group computing and context learning layer that contains the following components. The intelligent tagging module is responsible for assigning tags to the published public activities using heuristic rules. The context extraction module is in charge of inferring social/personal contexts (e.g., user location, preferences, who is together with) from raw sensory data. The group extraction and abstraction module is tasked to extract raw activity groups from activity logging data or third-party groups. In addition, it also distills logical groups from raw groups.

The sound-based activity recognition module is intended to predict the type of ongoing activities by analyzing the ambient sounds sensed from smartphones. The intimacy measurement module offers social structure or interaction-based metrics for intimacy calculation.

The fourth layer is the activity support layer, which meets the aforementioned needs on group activity organization.

6.4.5 Planned Group Activity Preparation

6.4.5.1 Group Activity Initiation

There are two types of group activities in terms of their openness, and we provide different support for them.

Private Activity: Users can initiate an activity through its category tag, venue, and time. A category tag can be selected by an initiator from a predefined tag set. To facilitate the selection of a venue, we divide the city map into 100 m × 100 m cells; each can be chosen as the activity venue. A third-party point of interest (PoI) dataset such as DBPedia [138] and LBSN check-ins [139] can be leveraged for fine-grained venue setting. As private activities are often associated with static groups, a list of group members for a private activity can be recommended at the initiation time based on ambient contexts.

Public Activity: To better communicate with the public, we need to learn from existing communication mediums. Posters are a popular and easy-to-use medium for public activity advertising. However, they suffer from problems such as spatiotemporal constraints and low dissemination speed [140]. Using chapter posters as the metaphor, we digitize this medium to address its weaknesses. Specifically, when initiating a public activity, people can publish a digital post to MobiGroup for sharing. The metadata (e.g., activity venue and category tag) for the activity post can be specified by the publisher. Users can browse the public activities from a map (e.g., by activity venues). They can interact with the posts and "save" the interested ones that they want to attend (referred to as "savers" to those posts).

6.4.5.2 Publicity Support for Public Activities

The publicity of planned activities is to advertise the activity posts to relevant people within a social community. In terms of the manner activity posts are acquired, there are two generic modes, i.e., pull and push [141]. Pull means that people can browse and query activity posts using prespecified category tags. By push, we compute the semantic tags and automatically circulate the recommended posts to potential interested people. Semantic tags are crowd-related knowledge and need to be learned from large-scale crowd-post interaction data. The methods are presented below.

Public Activity Sensing: To facilitate public activity publicity, it is important to record crowd interactions with the posts. The following information regarding the posts is kept:

Savers to a post: The people who interact with a post p (i.e., the "savers") are recorded, forming a post group $G(p)$.

Interaction time: It indicates the interaction time of each user to a post, i.e., when the post is saved by the user.

Social links: The social links among users are extracted from third-party SN services.

Feature Extraction: The following features are extracted from crowd-post interactions for semantic tag prediction.

Post group size: It denotes the number of "savers" to a post (i.e., the post group size), formulated as $|G(p)|$.

Temporal interaction context: The popularity of a hot post usually reaches a peak within a short period of time, similar to trending topics in Twitter. We divide 1 day into L time intervals. Let the number of savers to post p at the time interval T_i $(0 < i \leq L)$ of date d be $N(p,T_i,d)$ and the average number of saves to any posts at the same time interval during the past few days (we used 1 week in the current study, because the posts published earlier show little relation with the current post [142]) be $AvgN(T_i)$. We formally define the interaction index as follows:

$$II(p, T_i, d) = N(p, T_i, d)/AvgN(T_i), \tag{6.9}$$

where we use II to measure activity post p's interaction index during T_i. It is a dynamic feature and we use the maximum II in the interaction history of p as its current hot index HI, as formulated below:

$$HI(p) = Max\left(\{II(p, T_i, d) : d_{pub}(p) \leq d < d_{now}; 0 \leq i \leq L\}\right), \tag{6.10}$$

where $d_{pub}(p)$ refers to the publishing date of p, now refers to the current time, and d_{now} returns the date of the current time.

Post group density: We use the link density [143] to characterize the density of social connections among the savers of a post group, which is the fraction of the existing numbers of connections of savers in the post group against all possible connections. The metric is given as

$$LinkD(G(p)) = \frac{2 \times NumOfConns(G(p))}{|G(p)| \times (|G(p)| - 1)}, \tag{6.11}$$

where NumOfConns denotes the number of existing connections among the members of the post group $G(p)$.

Heuristic-Based Activity Advertising: We use heuristic rules to infer semantic tags. Specifically, we use the group size and the temporal interaction context to characterize a "hot" post. In the interaction history with an activity post p, if $|G(p)|$ is among the top $k\%$ of all the "active" posts and HI exceeds a threshold Th_{hi} at a certain time, it is tagged "hot." A post will expire when it passes the starting point of its A_Time, and it is viewed as an active post before expiration, formulated as

$$\text{Active}(p) \leftarrow \text{Now} < A\_Time(i). \tag{6.12}$$

The whole rule is defined as

$$\text{Hot}(p) \leftarrow \text{HI}(p) \geq \text{Th}_{hi} \Lambda p \in \{i|\text{Top}_k(|G(i), \{m|\text{Active}(m)\}) = 1\}, \tag{6.13}$$

where the Top function determines whether $|G(i)|$ is within the top $k\%$ of all active posts.

We use the density of a post group to determine whether a given activity post should be labeled as "social." For example, we check the group density of a certain activity's savers, and if the density value is above a predefined threshold Th_{ss}, the post is tagged "social." As mentioned in the above section, "social" posts are suggested to existing online group(s) if a proportion of members from an online group save the post. By applying a community detection method over the SN of users (can be obtained from third-party services), implicit groups (forming a group set called DetGroups) among them can be obtained. We can locate the group that contains the most people of a post group and recommend the associated post to the other members of this group. We formulate it as

$$G_{\text{social}} = \text{MaxGroup}(G(p), \text{DetGroups}), \text{ if } \text{LinkD}(G(p)) > T\text{ss}, \tag{6.14}$$

$$\text{Re } c(i,p) \leftarrow \forall i \in G_{\text{social}}, i \notin G(p). \tag{6.15}$$

where MaxGroup returns the group from DetGroups that contains the most people of $G(p)$, and Rec(i,p) denotes that p should be recommended to user i.

6.4.5.3 Group Suggestion for Private Activity Preparation

Private activities usually show regular patterns and relatively fixed group members, based on which we design an approach to supporting private activity group formation. As shown in Fig. 6.20, the approach consists of two modules: off-line group computing and online group suggestion. The prior module analyzes the relations among raw groups and merges them into logical groups, which can work off-line with historical social activity data. Online group suggestion leverages real-time contexts for group filtering and ranking.

Social Graph Model: To study the grouping patterns or the rules of private activities, we first build a model to characterize the network of private activity

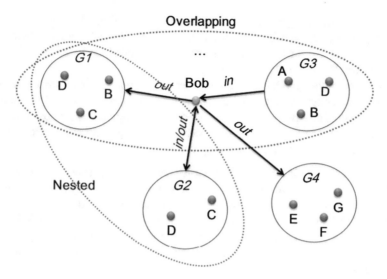

Fig. 6.21 Group activity graph example

participation. The social graph model is then proposed. In this model, the edges are formed by initiating or participating activities. We employ the egocentric network method used in ref. 107, in which a message sent by a user to a group of contacts is regarded as one that forms a single edge (a hyperedge). The edge is directed, represented as either in or out edges, corresponding to user-initiated or user-participated activities, respectively. All user-initiated group activities form a set called GAout, while all user-participated activities form the GAin set. The nodes to each hyperedge refer to the members of a certain activity and they form a raw group.

Figure 6.21 is an example of Bob's social graph, where four raw groups are involved (e.g., G1 to G4). For instance, G1 is formed by an activity which Bob initiates and B–D are members. G2 is associated with several activities, where Bob is either the initiator (the out edge) or a participant (the in edge).

Private Group Computing: People can participate in various group activities, and different activities usually link different group instances. This relation results in a large number of groups in the initial social graph. This is a challenge for initiating private group activities, considering that there are many overlapping or nested groups. For instance, A can be involved in both a sport team and a project team, and the two groups overlap. This is also obvious from the social graph model. As shown in Fig. 6.21, the social graph of Bob consists of overlapping (e.g., groups G1 and G3) and nested groups (e.g., groups G1 and G2). G4 is more specific and has little relation with the other three groups. Regarding different themes of private group activities, a user's social ties can be clustered into different subcommunities, which is called the implicit grouping pattern. For example, A usually has lunch together with B, C, and D, except for 1 day that B does not come for some reason;

this results in two raw groups: B, C, and D and C and D. From the implicit grouping pattern perspective, the two group instances should be merged as a logical group: B, C, and D. It is not difficult to derive that recommendations based on raw group instances can result in missing data or errors.

To discover the implicit grouping patterns and improve the performance for private group suggestion, the group abstraction (GA) concept is used to eliminate minor subsets of groups by merging highly nested or overlapping groups into logical groups. We refer to the merging of nested groups as group subsumption and the merging of overlapping groups as group combination.

1. Group subsumption: Given two nested groups to a user u, G1 and G2 (G2 $\subset$ G1). The two groups can be subsumed if they are highly nested. We refer to MacLean et al.'s information leak metric for group nesting evaluation [106]. The value of information leak is determined by two factors: similarity of the two groups and ratio of the number of activities held by each group. We thus define a new parameter subrate to measure whether two groups can be subsumed, expressed as

$$\text{subrate}(G1, G2) = \frac{\mid G1 \mid - \mid G2 \mid}{\mid G1 \mid} \times \frac{\text{num}(G2)}{\text{num}(G1)}, \text{when}(G2 \subset G1), \quad (6.16)$$

where $|Gi|$ refers to the number of members of group Gi, and $(|G1| - |G2|)/|G1|$ defines the similarity of two groups; num(Gi) refers to the number of activities held by Gi. Note that both in and out relations between u and Gi are used. Suppose $G1 = \{B, C, D\}$, $G2 = \{B, C\}$, and 100 and 5 records are relevant to $G1$ and $G2$ in ARC, respectively; then we have subrate($G1$, $G2$) = (3–2)/ 3 $\times$ (5/100) = 1/60. If the value is below a predefined threshold (Th_{sub}), the two groups can be subsumed.

2. Group combination: Two overlapping groups can be combined if they are very similar. To measure the similarity between two groups, we use the Jaccard metric, expressed below. The two groups can be combined if their similarity exceeds a threshold (Thcom):

$$\text{comrate}(G1, G2) = \frac{\mid G1 \cap G2 \mid}{\mid G1 \cup G2 \mid}, \text{when overlap}(G1, G2). \quad (6.17)$$

GA results in a set of logical groups, which facilitates the management of groups (considering that the number of raw groups can be rather large).

Context-Aware Group Filtering: One basic principle for group cueing is to suggest relevant groups according to user needs. Various contexts can be used to filter out irrelevant logical groups when users initialize activities.

Spatial-temporal contexts: It includes the location and time context regarding the activity being initiated. Location can be obtained by in-phone GPS positioning or Wi-Fi indoor positioning techniques [144]. The initiation time context is specified as one of the four logical periods, such as morning (6:00 to 11:00) and noon (11:00 to 13:00).

WithWhom: Nearby friends who are often coinitiators or members of an activity. We use WithWhom (i) to indicate that a number of i contacts are together with the initiator. This context can be obtained using the Bluetooth ID of user mobile phones.

Tag: A category tag specified by the initiator often shows the type of the activity being organized.

The rule for group filtering is performed in this way: for each context C_i that is obtained when a new activity is organized, a logical group G_j is considered irrelevant and thus filtered if it has no historical record that matches C_i.

Group Ranking: After context-aware filtering, there can be still one more group remaining. To suggest the most relevant group to the user, we developed a group affinity ranking method to rank the remaining groups by the strength of the link between a user and the user's logical groups. We employ the method used in ref. 107, which was originally used for link-strength measurement in email networks. The link strength between two entities is computed based on their interaction history.

In MobiGroup, it is measured based on the group activity history between a user and a logical group. In addition to interaction frequency (i.e., the number of associated activities), recency and user role are also considered in group affinity ranking. Human relationships are dynamic over time, and we use recency to denote that the recent data is more important than the old data. For user role, the activities in which the user is the initiator (i.e., the out relation) are more important than those in which the user is merely a participant (i.e., the in relation). We define the affinity rank between user u and the logical group G_k as affRank(u, G_k), which can be calculated as

$$\text{affRank}(u, G_k) = w_{\text{out}} \sum_{A \in GA_{\text{out}} \wedge G(A_i) = G_k} \left(\frac{1}{2}\right)^{d_{\text{now}} - d_{\text{pub}}(A_i)}$$
$$+ w_{\text{in}} \sum_{A \in GA_{\text{in}} \wedge G(A_i) = G_k} \left(\frac{1}{2}\right)^{d_{\text{now}} - d_{\text{pub}}(A_i)}, \qquad (6.18)$$

where ω_{out} and ω_{in} represent the weights of the user roles in social activities, with the former being larger to represent the importance of initiator roles. We empirically use 1.5 and 1.0 in the current implementation. $d_{\text{pub}}(A_i)$ refers to the initiation date of activity A_i. Given u, the implicit group with the highest rank will be recommended.

We use an example to demonstrate the group affinity ranking method. As shown in Table 6.4, for user u and a logical group G1, there are three historical activities (A1–A3) associated with them. We can also find the metadata to each activity, such as user role (in/out relation) and activity date. To illustrate our method, three cases are given. The only difference between cases 1 and 2 is that the user role to A2 changes (from out to in relation). We find that the rank value of case 1 is higher than that of case 2, which indicates that the out relation (i.e., the activity initiator role) is weighted higher in our method. Similarly, the only difference between cases 2 and 3 is that the activity date changes. We find that the rank value of case 2 is higher than that of case 3, which reveals the effect of the recency factor.

Table 6.4 Example for affinity-based group ranking

Case ID	Associated activities (in/out, date)	affRank Value
1	A1(out, Nov. 15), A2(out, Nov. 16), A3(in, Nov. 14)	$1.5*(0.5\wedge2 + 0.5) + 0.5\wedge3 = 1.25$
2	A1(out, Nov. 15), A2(in, Nov. 16), A3(in, Nov. 14)	$1.5*(0.5\wedge2) + (0.5 + 0.5\wedge3) = 1$
3	A1(out, Nov. 15), A2(in, Nov. 15), A3(in, Nov. 14)	$1.5* (0.5\wedge2) + (0.5\wedge2 + 0.5\wedge3) = 0.75$

6.4.6 Running Activity Recognition and Suggestion

In addition to augmenting the preparation process for group activities, we also provide support for running activities. Specifically, if users permit sharing their current activities, MobiGroup can recommend the activities to relevant people based on social/physical contexts. It is particularly helpful for unplanned activities, which do not have the preparation process.

6.4.6.1 Ambient Sound-Based Activity Recognition

To suggest a running activity to other people, it is important to first identify the category of the activity. Previous studies on human activity recognition [111, 145] mostly focused on individual activities. Some explore group activity recognition using mainly computer vision techniques [112, 146]. Nevertheless, the vision-based method is computationally intensive and intrusive. We have developed a less intrusive approach to group activity understanding, based on the analysis of sensed ambient sounds from smartphones. There have been studies on ambient sound-based group identification [112, 147] or place categorization [148], and our work is designed for a different purpose.

To reduce the cost when running on resource-constrained devices, a two-stage recognition method is used. Currently, our work used the smart campus as the test bed and identified three most common activity types, namely talk, relaxation, and other activities. The relaxation activities are associated with music events (e.g., in a party, in a shopping mall). The data recognized as other activities will be further processed in the second stage. We identify three popular activities that university students undertake on a daily basis: street roaming, sporting, and in a quiet environment (e.g., in a library). The recognition algorithm is given below.

Preprocessing: The raw audio data must be preprocessed because it cannot be recognized directly. Ambient sound in the form of audio streams is captured by the mobile phone's microphone sensor. The audio stream data is segmented into frames for feature extraction. The sampling rate is 8 kHz, and each frame length is set to 64 ms to enable lower duty cycle on the phone. In speech recognition, researchers often set the frame overlap between two continuous frames to enhance smooth analysis. The frame overlap is generally set to 32 or 64 ms. As mobile devices have limited resources, we do not set frame overlap in this study. To compensate the high-frequency part that is suppressed during the sound production process, the input signals should be pre-emphasized [149]. Finally, because the change of audio signal in the time domain is difficult to distinguish, the frames are processed using the fast Fourier transformation method for further analysis.

First-Stage Activity Categorization: According to the requirements of MobiGroup, three features are selected, namely zero-crossing rate (ZCR) [45, 150], spectral flux [151], and Haar [152]. Each selected feature is useful for distinguishing the associated acoustic events. For instance, comparing with music, speech signals often show a higher variation in ZCR [45]. Haar-like filtering is traditionally used for image processing, but it is also effective for ambient sound recognition [152]. Based on the features, we build a classifier using the J48 decision tree algorithm, which is efficient when running on resource-constrained devices [152].

Second-Stage Activity Recognition: For the second-stage recognition, the goal is to classify the ambient sounds identified as "other" in the first stage into three activity types. The 24-D MFCC features [149] are used, and the DTW algorithm [153] is applied to classify the three types of activities. MFCC features are more powerful but computationally intensive. We avoid using them over the large number of raw audio clips in the first stage.

6.4.6.2   Context-Aware Running Activity Recommendation

When a user shares his/her status online, it indicates that he/she wants to have opportunistic participants to join his/her activities. The host user may have different preferences or constraints for the participants. As depicted in the above section, we attempt to address two varied constraints: one is to invite close friends (the SI), and the other is to suggest the ones that have frequent interaction recently (the DI). Several other contexts should also be considered, such as preferences of friends, and the distance between the activity venue and the host user's location. This is achieved through the fusion of social interaction data gleaned from online and off-line communities. Online community data are used to characterize the social intimacy (static or dynamic) between users, while the off-line data refers to the user contexts in the real world. In other words, we are trying to generate hybrid groups based on online third-party data and associated contexts in real-world scenarios.

Preference: We use P (u_i, A_j) to denote user $u_i's$ preference degree to activity category A_j. It can be calculated based on the user's historical activity records in ARC. Since preferences can change over time, we use the observation data of the last 30 days for preference learning. It is formulated as

$$P(u_i, A_i) = \frac{\text{NumAct}(A_j, u_i, \text{ARC}(d))}{\text{NumTotalAct}(u_i, \text{ARC}(d))},$$ (6.19)

where ARC(d) refers to the subset of historical data over the past d days, set to 30 in current study; NumAct and NumTotalAct refer to the number of activity instances a user participates for activity category A_j, and the total number of activity instances the user participates, individually.

Distance: It denotes the distance between the venue of an activity and a user. In general, within a certain range of the distance Th_d, the impact of the distance on user's choice is low and the distance factor changes slowly. Beyond the threshold, though, the distance will become more and more important as its value increases. Based on this consideration, we construct the distance parameter Dis as

$$\text{Dis} = \begin{cases} \left(\sqrt[Th_d]{2} \right)^{\text{distance}}, & \text{distance} < Th_d \\ \left(\sqrt[Th_d]{2} \right)^{\text{distance}} \cdot \left(\sqrt{2} \right)^{\text{distance}-Th_d}, & \text{distance} \geq Th_d \end{cases}, \qquad (6.20)$$

where Th_d is set to 3 km in the current study. If the distance is over 3 km, the growth rate of Dis will become increasingly higher.

Dynamic Intimacy: DI supports the formation of interaction-based groups using third-party data. It is measured by the interaction frequency in OSNs between a user and his/her friends, based on the interaction data (e.g., comments, likes) on each other's post wall. DI characterizes the dynamic relation among users because the interaction frequency among them may change over time. We define the interaction factor of user u_i to u_j based on u_j's comments to u_i's posts during the past 30 days (considering the dynamic nature of interaction) as follows:

$$I_{\langle u_i,u_j \rangle} = \sum_{k=1}^{n} \left(\frac{1}{2} \right)^{\Delta t(k)} \cdot CR, \Delta t(k) \leq 30, \qquad (6.21)$$

where n is the total number of comments of u_i to u_j. $\Delta t(k)$ denotes the recency, i.e., the number of days between the comment date of k and the current date. CR denotes the ratio of user u_j's comments to u_j over all the comments to u_j. For the DI between two users (u_i, u_j), we should consider both directions of interaction: u_i to u_j and u_j to u_i. The DI of user u_i to u_j is defined as

$$DI_{\langle u_i,u_j \rangle} = w_1 \cdot I_{\langle u_i,u_j \rangle} + w_2 \cdot I_{\langle u_j,u_i \rangle}. \qquad (6.22)$$

We consider that a user's comments to his/her friends are more important for recommending activities to the user; therefore, it should have a higher weight. In this study, the two weights w_1 and w_2 are set to 1.0 and 0.5.

Static Intimacy: It refers to the relatively long-term or static relationship among users. Two measures—mutual intimacy (MI) and coverage intimacy (CI)—are calculated from the social tie graph for SI measurement.

According to Mitchell [154], if there are more mutual friends between two users, and there are more social ties among the mutual friends, the two users are highly likely close friends. We have proposed a method to calculate static user intimacy

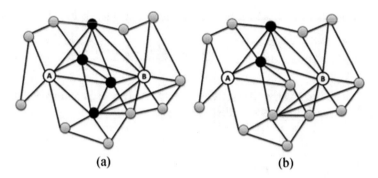

(a)   (b)

Fig. 6.22 Examples of social tie connection between users A and B

based on this finding. As shown in Fig. 6.22a, the gray nodes represent a friend of A
but not a friend of B, or a friend of B but not a friend of A, whereas the black nodes
represent the mutual friends between A and B. According to the relationship between
common friends, the MI ratio between u_i and u_j can be defined as

$$\mathrm{MI}\left(u_i, u_j\right) = \frac{2 * e}{M * (M - 1)},\tag{6.23}$$

where e denotes the actual number of edges among mutual friends, and M refers to
the number of mutual friends of two users. For example, the MI ratio between users
A and B (as shown in Fig. 6.22a) equals to 5/6.

If two users only have two mutual friends and there is one edge between the two
friends, the value of MI has the maximum value (i.e., MI = 1), as shown in
Fig. 6.22b. However, we cannot derive that the two users are very close. For
example, the relationship between A and B seems closer in Fig. 6.22a than in
Fig. 6.22b, but the MI value does not indicate this. In addition to MI, we define
another parameter, i.e., the CI ratio for SI measurement. CI measures the proportion
of the number of mutual friends between u_i and u_j to the total number of u_{ji}s friends,
which can be formulated as

$$\mathrm{CI}\left(u_i, u_j\right) = \frac{M}{F(u_i)},\tag{6.24}$$

where $F(u_i)$ denotes the total number of $u_i's$ friends.

We define the SI of user u_i to u_j based on the above two intimacy ratios,
formulated as

$$\mathrm{SI} < u_i, u_j >= MI\left(u_i, u_j\right) \cdot \mathrm{CI}\left(u_i, u_j\right).\tag{6.25}$$

Accordingly, the SI of A to B in Fig. 6.22a, b is calculated as 10/21 (5/6 × 4/7)
and 1/3 (1 × 2/6), respectively. It indicates that the SI between A and B in Fig. 6.22a
is higher than that of Fig. 6.22b.

Finally, the ranking score for recommending $u'_j s$ activity A_k to u_i is defined as

$$\text{Rank}_{\text{DI/SI}}(u_i, u_j, A_k) = \alpha * \frac{P(u_i, a_m) \cdot (DI_{<u_i,u_j>} \text{ or } SI_{<u_i,u_j>})}{Dis}, \qquad (6.26)$$

where α is a constant value. The activity information can be recommended to the list of users in terms of the order of their ranking scores.

6.4.7 Performance Evaluation

To evaluate the performance of the proposed methods, we implemented a prototype of the MobiGroup system on the Android platform. Based on the prototype, we designed a set of experiments to test the system's performance from two aspects, i.e., public activity advertising and private activity suggestion. For more information about the experimental setup, please refer to the article [155].

6.4.7.1 Public Activity Advertising

The first experiment investigated whether users would benefit from our heuristic-based approach to public activity advertising. We used the dataset of activity posters in our university collected from 38 student volunteers [140]. Users' average rating scores on recommendations by each method are shown in Fig. 6.23, based on which we can obtain the following findings. First, people prefer the "hot" and "social" posts

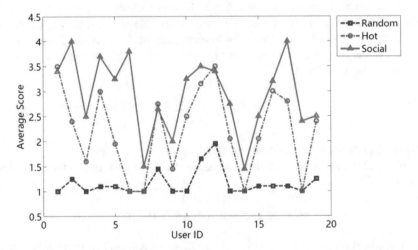

Fig. 6.23 Results for tagging with heuristic rules

Table 6.5 Effects of contexts to group suggestion

Contexts	Precision (%)	Recall (%)
AR (AffRank) + WithWhom(1) [117]	35	40
MobiGroup—Context	29	31
MobiGroup—GA	40	44
Time + I-Loc	57	75
Time + I-Loc + tag	70	81
Time + I-Loc + WithWhom(1)	70	93
Time + I-Loc + WithWhom(2)	84	98

suggested by our method, given that these posts normally have higher rating than the randomly chosen ones. Second, the average scores of "social" posts are higher than "hot" recommendations, indicating that people in the same detected group do share some common interests, and group preference-based recommendation is more effective than item popularity-based recommendation in loosely linked communities.

6.4.7.2   Group-Aware Private Activity Suggestion

The second experiment was designed to investigate whether the group computing and context-based methods are effective to support private activity group formation and event suggestion, and the Friend Suggest (FriSug) method [107] was used as the baseline method for comparison.

First, a major difference between the proposed MobiGroup system and other group tools is that MobiGroup leverages rich contexts derived by smartphone sensing to provide more relevant group suggestions. Specifically, we chose four types of contexts to filter irrelevant groups, with Time and I-Loc (initiation place) as the basic criteria and Tag, WithWhom(1), and WithWhom(2) as additional ones. Experimental results are given in Table 6.5, which indicates that rich contextual information is able to enhance the recommendation performance and that the WithWhom context performs better than the Tag context.

Meanwhile, group abstraction (GA) contributes to the management of groups as well, which can be used to eliminate noise in group detection and merge relevant raw groups into bigger logical units. Results in Table 6.5 suggest that GA can characterize the social graph of a user and enhance the performance of private group suggestion. To sum up, based on spatiotemporal contexts and GA, the proposed MobiGroup system significantly outperforms the FriSug system [107], e.g., the precision increases from 35 to 70%, indicating that the two factors are useful for group suggestion. For more details about the experimental results, please refer to the article [155].

6.5   Predicting Activity Attendance in Mobile Social Networks[5]

6.5.1   Introduction

With the proliferation of event and activity-based applications such as Facebook Events, Google Events, Meetup, and Douban Events4, it has become possible for users to expand their online interactions to off-line activities. People can propose and attend a variety of off-line social activities through these online services, which can further promote face-to-face social interactions. This kind of social media is called event-based social networks (EBSNs) [156]. Although there have been studies on EBSNs, very few attempts have been paid on behavior prediction. Understanding the collective dynamics of user participation in events is crucial to better understand social networks in both the physical and cyber worlds and to provide critical insights that help personalized event recommendation and targeted advertising. In contrast to traditional social network services (SNS), user behavior in EBSNs is predominantly driven by off-line activities and highly influenced by a set of unique factors, such as spatiotemporal constraints and special social relationships (host and attendee). Therefore, these properties of social activities play an important role in behavior prediction in EBSNs, making it different from the online behavior prediction in SNS. Figure 6.24 shows the main elements of activities in Douban Events, a popular EBSNs in China, described as follows.

Location: where the activity will be held. Usually, an activity is held at a convenient and popular place.

Time: when the activity will start and end. In fact, the distribution of time highly depends on the content of an activity.

Attendees: users who click the "I want to attend" button to show their intention to attend the event.

Host: the user or site who will hold the event and be responsible for off-line activity organization.

Content: the detailed information of an activity, which includes three aspects: category, title, and description. Categories could be music, film, sport, party, travel, exhibition, drama, and so on.

Based on these elements, we are able to predict a user's activity attendance using different methods. However, how to comprehensively combine the heterogeneous information in a systematic way remains an open issue. In this chapter, we formulate and parameterize the above elements into a multifactor model, which mainly consists

---

[5]Part of this section is based on a previous work: R. Du, Z. Yu, T. Mei, Z.T. Wang, Z. Wang, B, Guo. Predicting activity attendance in event-based social networks: content, context and social influence, Proceedings of the 2014 ACM International Joint Conference on Pervasive and Ubiquitous Computing, Seattle, USA, September, 2014, pp. 425–434

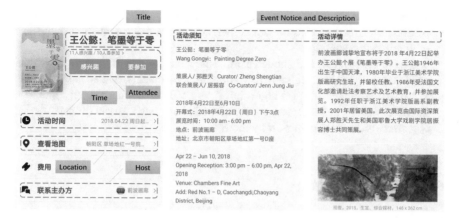

Fig. 6.24 An example of an activity on Douban Events with five key elements: location, time, attendees, host, and content

of three different factors: (1) *Content*. One key factor that determines whether a user will attend a certain activity is its content. We use content preference to represent a user's preference for different activities. (2) *Context*. The context in our method refers to spatial context and temporal context. Specifically, users may have different time and location preferences while attending different off-line activities. For example, if a user is a student, the possibility for him/her to attend an activity held during a weekday would be low, even if he/she likes the activity. Similarly, if a user is far away from the activity location, his/her willingness to attend would be low. (3) *Social influence*. In EBSNs, users follow each other based on common interests. One's willingness to attend a certain activity could be impacted by his/her social relationships with the host and other attendees. However, due to the time and location constraints of activities, the host usually plays a much more significant role in user's attendance decision than ordinary followers. The reason is that, on the one hand, the host can recommend activities to its followers and, on the other, if a user often attends activities hosted by an influential host, the possibility for attending other events organized by this host will also be high.

Based on these factors and the multifactor model, we propose a novel method, named singular vector decomposition (SVD) with multifactor neighborhood (SVD-MFN), to predict activity attendance in EBSNs. The method is based on SVD and is capable of integrating different features into a multifactor model as the neighborhood, which fully takes advantage of both the SVD and multifactor model. The main contributions can be summarized as follows:

- We propose to investigate the prediction of activity attendance in the emerging EBSNs, which is, to the best of our knowledge, one of the first attempts in the area of ubiquitous computing.
- We have discovered and modeled three key factors that influence individual behavior, e.g., content preference, spatiotemporal contexts, and social influence.
- We have developed a novel algorithm (i.e., SVD-MFN) to optimally integrate multiple heterogeneous factors we mentioned above into a single framework that outperforms the state-of-the-art methods.

6.5.2   Related Work

We briefly review related work, which can be classified into three categories. The first category is research on understanding relationship between online and off-line social interactions in EBSNs. Liu et al. proposed the concept of EBSNs and focused on community detection using both online and off-line social links [156]. Pramanik et al. [157] took a step to quantify the success of Meetup groups in EBSNs, and achieved high accuracy in predicting success of Meetup groups. Han et al. sought to gain insights into user behavior for attending off-line events based on Douban Events [158]. By studying the events in Douban, they present results linked to the event properties, user behavior of participants of an event, and social influence on an event, which help us better understand what affects user behavior during events. Xu et al. conducted a quantitative analysis that revealed the relationship between online following behavior and characteristics of real-world events [159]. The difference between our work and this line of research is that we move one step further to leverage the unique characteristics of EBSNs to predict whether a user will attend an activity or not.

The second category includes research into activity or event prediction. There are some approaches towards activity recommendation. For example, in the Pittsburgh area, a cultural event recommender was built around trust relations [160]. Cao et al. [161] proposed a multifeature-based recommendation system in EBSNs, which aims to recommend users with social events according to their preferences. Liu et al. [162] studied how to exploit diverse relationships in an EBSN as well as individual history preferences to recommend preferred events. Li et al. [163] proposed a random walk-based event recommendation algorithm. The recommender system for academic events focused more on social network analysis (SNA) combined with collaborative filtering (CF) [164]. Cornelis et al. developed a hybrid event recommendation approach where both CF and content-based algorithms were employed [165]. Daly et al. established an interesting event management service that considers the location of events when making recommendations [166]. They found that event attendees are sometimes from nearby locations and proposed a location-based method to recommend local events to targeted users. Minkov et al. presented a collaborative approach for event recommendation which outperforms approaches that only consider content information [167]. Sang and Zhuang et al. explored using temporal and spatial context for entity recommendation [168, 169]. However, the existing research only considers one or two aspects of events in EBSNs, and none of them have developed a comprehensive yet systematic method to combine different and heterogeneous information.

The third category focuses on location-based social networks, which also contain both online and off-line social interactions [170, 171]. Although adjacent check-ins may indicate implicit social interactions and social ties [172], the check-in data is usually too sporadic to represent human behavior [173]. Crandall et al. examined the geographical features to infer social ties [174]. Similarly, Zhuang et al. formalized the problem of predicting geographic coincidences in ephemeral social networks and used a factor graph model to predict the possibility of two users meeting in future [175]. Yi et al. [176] also proposed a framework that aims to predict future encounters between different users, and supports music sharing according to their preferences. However, neither of these prediction methods is event driven. Additionally, while the above research mainly used spatial information for prediction, the "off-line" (social events) features considered in this chapter contain multiple specific characteristics of the activity, such as content preference, spatial and temporal context, and social influence.

6.5.3   Problem Statement and System Overview

6.5.3.1   Problem Statement

In EBSNs, there is a list of activities $\{a_1, a_2, \ldots, a_m\}$ and a list of users $\{u_1, u_2, \ldots, u_n\}$. Let A_u denote all the activities that a specific user u has ever attended. User preference is usually extracted from historical behavior. We use $F_u(a)$ to represent the preference of user u for the activity a, which consists of the following five parts. $CP_u(a)$ is the content preference, $DP_u(a)$ is the distance preference in the spatial context, $WP_u(a)$ and $SP_u(a)$ are the day of the week and the hour of the day preferences in the temporal context, and $SP_u(a)$ represents social influence, which means the relationship of u and a's host. These factor pairs can be concluded as three macro aspects: content, context, and social influence. We can also get an activity neighbor set $NS_u(a)$ for each user-activity pair selected from A_u by combining the above three aspects, which means the nearest activities selected from user's attendance history. The activity attendance prediction task of a target user in EBSNs can be described as follows: given a user an attendance history A_u and the set of attendees U_a of an upcoming activity a, we are predicting whether the target user u will attend a or not. The prediction result is represented as r_{ua}, which is a binary value; that is, 1 means "attend" and 0 "ignore." In this chapter, we take both $F_u(a)$ and $N S_u(a)$ into account for effective activity attendance prediction. The key problems are listed as follows:

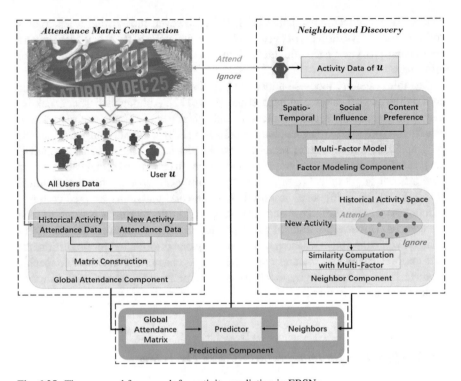

Fig. 6.25 The proposed framework for activity prediction in EBSNs

- How to evaluate contribution of different factors including $CP_u(a)$, $DP_u(a)$, $WP_u(a)$, $HP_u(a)$, and $SP_u(a)$? How to organize and fuse these factors in a single model, and then obtain $Fu(a)$?
- How to extract the neighbor set $NS_u(a)$ using the factor model?
- How to combine $F_u(a)$ and $NS_u(a)$ in the prediction of r_{ua}?

6.5.3.2   System Overview

Figure 6.25 shows an overview of our framework, which consists of three components: the attendance matrix construction component (left), the neighborhood discovery with multifactor component (right), and the prediction component (bottom).

Attendance Matrix Construction. On the one hand, given the target user u and the current attendees of an upcoming activity a, we can construct a binary attendance matrix for activity a. On the other hand, the historical attendance matrix can be built for all the users. Afterwards, based on matrix updating, we can get a global attendance matrix, which is one input for the prediction component.

Neighborhood Discovery with Multifactor. The second component is based on the target user's historical attendance, which is the key component in our framework.

We first extract features for each activity from three aspects: content preference, spatiotemporal context, and social influence. Then, considering the different influence of these three aspects, we propose a multifactor (MF) model using decision tree to evaluate their contributions. Based on this model, we can compute the similarity between each pair of events from the target user's perspective. Furthermore, we can discover the neighbor activities for the target user from the attendance history.

Prediction Component. To combine the previous parts into our system, we propose the SVD-MFN predictor in the prediction component. We will give the details in the SVD-MFN algorithm section.

6.5.4   Feature Modeling

In this section, we first elaborate the features used in our multifactor model from three macro aspects: content preference, spatiotemporal context, and social influence. The foundation of this model is a user study of the factors that influence activity attendance, which shows that users pay the most attention to these three aspects when they choose activities. Then, we present how different features are fused together.

6.5.4.1   Feature Extraction

The content of an activity plays a major role in determining the likelihood of a user participating in an event [167]. Therefore, the calculation of content similarity between the user's historical activities and the upcoming activity is a key step. A proper similarity measure can greatly influence the performance of our system. In the case of Douban Events, there are three elements for characterizing the activity content: category, title, and description, as shown in Fig. 6.25. We put these elements together as a whole text, and then define the content similarity between a pair of activities as their text similarity. Numerous studies have attempted to resolve the text similarity problem. One approach is to expand and enrich the keyword in the text with a search engine [177]. Another approach uses an external lexical database, such as WordNet, to mine the relationships among words [178]. Although there are a lot of words and their semantic relationships in the lexical database, the application scope of dictionary-based similarity computation is quite limited. The latent Dirichlet allocation (LDA) is a probabilistic topic model that can solve all the above problems [179].

LDA is based upon the idea that documents are mixtures of topics, where a topic is a probability distribution over words. We begin by removing stop words (i.e., punctuation) and short words (i.e., "of" and "and"), and then format the remaining text as the input of LDA. Afterwards, the formatted text is mapped to the subject space by using the Gibbs sampling. And then the Jensen-Shannon (JS) distance is often used to compute text similarity [180]. In the end, in order to get the best number of topics in LDA, we use a clustering method [177]. The process of generating a text with n words based on LDA can be described by a marginal distribution:

$$p(d) = \int_{\theta} \left(\prod_{i=1}^{n} \sum_{T(i)} p\left(w^{(i)} | T^{(i)}, \beta \right) P\left(T^{(i)} | \beta \right) \right) P(\theta | \alpha) d\theta, \qquad (6.27)$$

where $P(\theta|\alpha)$ is derived from Dirichlet distribution parameterized by α, and $P(W^{(i)}| T^{(i)}, \beta)$ is the probability of word $W^{(i)}$ under topic $T^{(i)}$ parameterized by β. The topic-word distribution $P(W^{(i)}|T^{(i)}, \beta)$ in Eq. (6.1) is an important factor for the implementation of LDA. We use the Gibb-s sampling method to extract topics from the corpus [181], and then adopt the result of sampling as the input of text similarity computation. A standard function to measure the divergence between two distributions p and q is the Kullback–Leibler (KL) divergence [180]:

$$D_{\mathrm{KL}}(p, q) = \sum_{j=1}^{T} p_j \log_2 \frac{p_j}{q_j}. \qquad (6.28)$$

The KL divergence is asymmetric but convenient to be applied in the JS divergence which has been proved symmetrized [180] and its value ranges from 0 to 1, formally defined as

$$\mathrm{JS}(p, q) = \frac{1}{2} \left[D_{\mathrm{KL}}\left(p, \frac{p+q}{2} \right) + D_{\mathrm{KL}}\left(q, \frac{p+q}{2} \right) \right]. \qquad (6.29)$$

As an example, consider the three sample activities which have been preprocessed:

$\alpha 1$: "drama British Shakespeare classic Macbeth."
$\alpha 2$: "British Shakespeare King Lear."
Applying LDA with topic number $Z = 2$ would yield topics to:
T1: "British Shakespeare."
T2: "drama classic."
Obviously, $\alpha 1$ would have a 50% membership in both topics, since it contains words from both topics to an equal degree, and activity $\alpha 2$ would have a 100% membership in T1 and T2, respectively. We could then represent each activity as a vector of their topic memberships:

$\alpha 1 = [0.5, 0.5]$
$\alpha 2 = [1.0, 0.0]$

where the first element in each vector corresponds to their distributions in topic T1 and the second element to distributions in T2, represented as $\theta^{(\alpha 1)}$ and $\theta^{(\alpha 2)}$. So, the JS divergence between $a1$ and $a2$ can be computed by taking $\theta^{(\alpha 1)}$ and $\theta^{(\alpha 2)}$ into Eq. (6.29), which equals to 0.31. Consequently, we could get the content similarity between $a1$ and $a2$ with JS divergence as

$$\text{Sim}(\alpha 1, \alpha 2) = 1 - JS\big(\theta^{(\alpha 1)}, \theta^{(\alpha 2)}\big), \tag{6.30}$$

which equals to 0.69 in the example.

Based on the content similarity of two activities, we can compute the user's content preference/interest. Specifically, we adopt the interest drift with forgetting mechanism in our work [182], and the key idea is explained below:

On the one hand, people's interest wanes as time goes by like memory. For example, an activity that was attended recently by a user should have a higher impact on the prediction of future behavior than an event that happened a long time ago. On the other hand, the forgetting speed slows down as the accumulated interests become more stable. Based on these two principles, we have constructed two interest models for different purposes by incorporating the forgetting mechanism, using short-term interest model (STIM) to represent the user's recent interests and long-term interest model (LTIM) to denote accumulated stable interests. The forgetting function is implemented to simulate the attenuation of the user's interests based on the following equation:

$$I(a, a_i) = \exp\left\{-\frac{\ln 2 \times (t_a - t_{a_i})}{hl}\right\}, \tag{6.31}$$

where the forgetting coefficient $I(a, a_i)$ denotes the degree that the original interest has declined, t_{ai} means the start date on an activity that the user has attended, t_a means the date of the future activity to be predicted, and hl denotes the half-life (in days) controlling the speed of forgetting. The larger hl is, the slower interests fade. When $t_a - t_{ai} = hl$, $I(a, a_i)$ descends to 1/2. For the short interest model, hl is a stable constant, and we set it as 90. For the long interest model, the half-life is not a constant any more. The user's interests usually become more stable over time. We use the forgetting formula to calculate coefficient $I(a, a_i)$ as

$$I(a, a_i) = \exp\left\{-\frac{\ln 2 \times (t_a - t_{a_i})}{hl_0 + d_{\text{acc}} \times s}\right\}, \tag{6.32}$$

where hl_0 represents an initial half-life value, and d_{acc} denotes how many days the original LTIM has evolved. Constant s reflects the impact of d_{acc} on the forgetting speed and we set it as 0.5 by experience. By involving factor $d_{\text{acc}} \times s$, the user's interests fall more slowly than original formula. To sum up, we use the following forgetting formula to calculate the coefficient:

$$I(a, a_i) = \begin{cases} e^{-\frac{\ln 2 \times (t_a - t_{a_i})}{hl_0}}, & (t_a - t_{a_i}) \le d_{\text{acc}} \\ e^{-\frac{\ln 2 \times (t_a - t_{a_i})}{hl_0 + d_{\text{acc}} \times s}}, & (t_a - t_{a_i}) > d_{\text{acc}} \end{cases}. \tag{6.33}$$

Consequently, for target user u, the content similarity between future activity a and past event a_i is $CS_u (a; a_i) = I(a, ai) \times Sim(a, a_i)$, where $Sim (a, a_i)$ means the content similarity. Considering both the content similarity between two activities and the attenuation degree of the user's interests, we define the content preference of user u to activity a as follows:

$$CPu(a) = \frac{\sum_{a_i \in A_u} I(a, a_i) \times Sim(a, a_i)}{\sum_{a_i \in Au} Sim(a, a_i)}. \qquad (6.34)$$

6.5.4.2   Spatial and Temporal Context

Temporal Context: The temporal context is important for activity prediction. On the one hand, as shown in Fig. 6.26a, human behavior shows strong daily and weekly periodical patterns. On the other hand, the start time of activities is also periodic, as shown in Fig. 6.26b.

1. *Day of Week Factor.* A user's daily life is usually weekly periodic, so we introduce the day of the week factor WPu (a) as the first temporal user preference as follows:

$$WSu(a, a_i) = \begin{cases} 1, wd(t_a) = wd(t_{a_i}) \\ 2, wd(t_a) \neq wd(t_{a_i}) \end{cases}, \qquad (6.35)$$

where wd (t_a) represents which day of the week the user attended activity a, and wd (t_a) 1, 2, 3, 4, 5, 6, 7 corresponds to the day of the week (i.e., Monday, Tuesday, Wednesday, . . ., Sunday). Thereby, u's day of week preference can be defined as

$$WPu(a) = \frac{\sum_{a_i \in A_u} WSu(a, a_i) \times Sim(a, a_i)}{\sum_{a_i \in A_u} Sim(a, a_i)}. \qquad (6.36)$$

2. *Hour of Day Factor.* In Fig. 6.26a, we found that a user's activity attendance in 1 day is also periodic, which can be explained as follows. Most users on Douban are either students or white collars. If an activity is held during study or work time, then these users are not likely to participate even though they are interested in the event. In contrast to the day of the week factor, we employ the Gauss formula instead of the binary formula to express the similarity in the hour level: $HS_u(a, a_i) = \exp\left\{ -\frac{(t_a - t_{a_i})^2}{2} \right\}$; thus we can express the hour of the day factor as

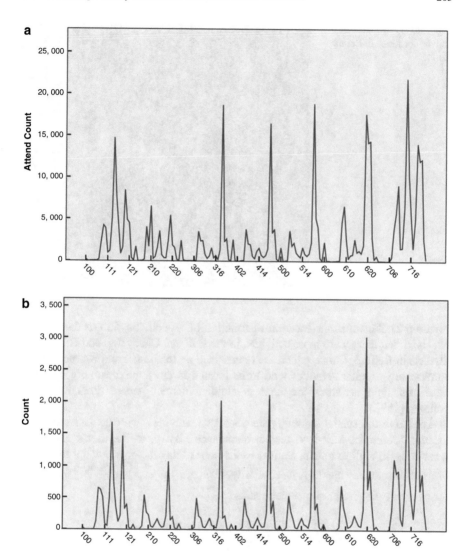

Fig. 6.26 Time histogram over hour of 1 week: (**a**) The user attendance time histogram over hour of the week. (**b**) The activity start time histogram over hour of the week

$$HP_u(a) = \frac{\sum_{a_i \in A_u} HS_u(a, a_i) \times Sim(a, a_i)}{\sum_{a_i \in A_u} Sim(a, a_i)}. \quad (6.37)$$

Spatial Context: We noticed that the likelihood of attending an activity decreases as the distance between the user's home location and the activity's location increases, which is not surprising and has been proven by numerous researchers [183]. Moreover, users have individual preference for locations. For example, if the transportation to a place is convenient, activities around it will be more popular.

Fig. 6.27 Distribution of activity locations in Beijing, China

Figure 6.27 illustrates the location distribution of all activities in our dataset. We observed that there were more activities in the Haidian, Chaoyang, and Dongcheng districts in Beijing, China, which are represented by the bigger and warmer circles. Additionally, similar activities tend to be located in the same areas. For example, most education activities are held in Haidian district, where there are many colleges [159].

To derive the spatial similarity between two activities, we calculate the user's location preference based on his/her attendance history, as the user's real home location is difficult to obtain. Similar to the hour of day factor, we adopt the Gauss formula to calculate the distance similarity, $DSu(a, a_i) = \exp\left\{ -\frac{\text{Distance}(a, a_i)^2}{2} \right\}$, and the user's spatial factor can be defined as

$$DP_u(a) = \frac{\sum_{a_i \in A_u} HS_u(a, a_i) \times \text{Sim}(a, a_i)}{\sum_{a_i \in A_u} \text{Sim}(a, a_i)}. \tag{6.38}$$

Social Influence: Social friendship is beneficial for event prediction and recommendation [184], which is a key factor motivating users to participate in social events [185]. We define two types of social relationships between the user and the host. The first type is the following relationship. In Douban Events, the activity host is allowed to send invitations to his/her followers. Therefore, if user u follows a host, he/she is likely to be more willing to attend the activities organized by this host once he/she is invited. The second type is preferring relationship. For example, a user may have attended a lot of sports events organized by one host who runs a gym, because the user is interested in this host or he/she is just a member of the gym. Therefore, he/she is more likely to attend future sports activities organized by this gym host whether he/she follows the host or not. Based on the above description, we define the social similarity between an upcoming event a and a past event a_i as

$$SS_u(a, a_i) = S_1(u, H(a)) * \delta + S_2(H(a), H(a_i)) * (1 - \delta). \tag{6.39}$$

where $H(a)$ means the host of a, S_1 means the following relationship, and S_2 means the preferring relationship. If u follows $H(a)$, $S_1(u, H(a)) = 1$; otherwise $S_1(u, H(a)) = 0$. The same goes for preferring relationship: for two activities a and a_i organized by the same host, if $H(a) = H(a_i)$, then $S_2(H(a), H(a_i)) = 1$; otherwise it equals 0. We set the parameter δ as 0.5, which means that these two types of relationships have the same weights. Consequently, the user's social influence is defined as

$$SP_u(a) = \frac{\sum_{a_i \in A_u} HS_u(a, a_i) \times Sim(a, a_i)}{\sum_{a_i \in A_u} Sim(a, a_i)}. \tag{6.40}$$

Now, we have extracted similarity features from three different aspects, which need to be fused together to enable effective activity prediction.

Feature Fusion: As different features have different impacts on user preferences, the challenge is to evaluate their significance and then fuse them together. In EBSNs, a user's rating for an activity is binary (i.e., 1 or 0), which means that he/she attends or ignores an event. Therefore, we can transform activity attendance prediction into a classification issue. Then classification algorithms can be used to deal with our problem. Based on the result of classification, we first get the contribution weights of different features and then combine the features linearly. The total similarity between an upcoming activity a and a past event a_i for the target user u is

$$Sim_u(a, a_i) = \alpha * CS_u(a, a_i) + \beta * WS_u(a, a_i) + \gamma * HS_u(a, a_i) + \delta \\ * DS_u(a, a_i) + \varepsilon * SS_u(a, a_i). \tag{6.41}$$

where α, β, γ, δ, and ε represent the weights of different features, respectively. In detail, α means the whole content preference weight; β and γ are the day of week and hour of day weights in the temporal context; δ is for the spatial context and ε denotes the weight of the social influence feature and their value ranges from 0 to 1.

6.5.4.3   SVD-MFN Algorithm

To effectively leverage different features for activity attendance prediction, we proposed a novel singular value decomposition with multifactor neighborhood (SVD-MFN) algorithm, which is based on SVD. Thereby, in this section we first introduce SVD and then explain the algorithm in detail.

Matrix Factorization: Singular value decomposition (SVD) is a well-known matrix factorization technique that addresses the problems of synonymy, polysemy, sparsity, and scalability for large datasets [186]. It has been proved in different areas, such as advertising [187] and e-commerce [188]. The basic idea of matrix decomposition in SVD is applicable to our system, because it delivers good results and is easy to tune and customize. In its basic form, every user u and activity a is associated with vectors as p_u, $q_a \in R^d$. The vectors p_u and q_a are generally referred to as d-dimensional latent user and activity factors, respectively. Based on these definitions, user u's attendance at activity a is predicted via the following equation: $r_{ua} = p_u{}^T q_a$. We use a logistic function to transform the preference score into the interval (0, 1), and set 0.5 as the threshold; that is, if it is bigger than 0.5 the user will attend; otherwise the user will ignore the event. In this work, parameters are generally learned by solving the following regularized least squares problem:

$$\min_{p*, q*} \sum_{(u,a) \in R} \left(r_{ua} - p_u{}^T q_a \right)^2 + \lambda \left(\sum_u \|p_u\|^2 + \sum_a \|q_a\|^2 \right), \qquad (6.42)$$

where the constant λ is a parameter determining the extent of regularization, which is set as 0.01. As we use the gradient descent learning algorithm, the training time grows linearly with the value of $|R|$. Therefore, the running time of each user-activity pair is $O(1)$.

SVD-MFN: Integrating Multifactor Neighborhood into SVD: Traditional neighborhood methods focus on computing the relationships between activities or, alternatively, users. While they are effective at detecting localized relationships and performing predictions on a few similar neighbors, they may fail to work when there is no or few observed ratings within the neighborhood of limited size. In contrast, the latent factor model in SVD is effective at capturing global information and has much better generalization capability due to its ability to represent users/activities more comprehensively. The neighborhood-based prediction model can be combined with the matrix factorization model as follows:

$$\tilde{r}_{ua} = p_u{}^T q_a + |N(u, a; k)|^{-\frac{1}{2}} \sum_{i \in N(u,a;k)} w_{ai} (r_{ui} - \bar{r}_u), \qquad (6.43)$$

where $\bar{r}_u$ means the average rate for user u and $N(u, a; k)$ consists of all the activities that are selected as the k-nearest neighbors of activity a attended by u. The parameter k refers to the number of neighbors. Specifically, we treat w_{ai} as free parameters which are learned together along with the matrix factorization model parameters. During computation, we only need to store and update the parameters for k-nearest neighbors of each activity instead of all the activity pairs. We set $N(u, a; k)$ as the k-nearest neighbors determined by the similarity measure $\text{Sim}_u(a, a_i)$, which integrates all the extracted factors.

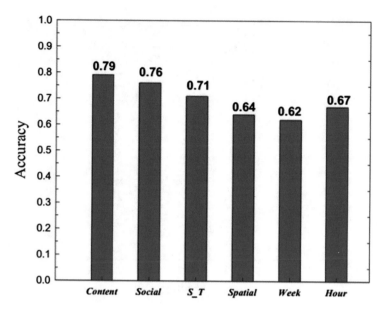

Fig. 6.28 Prediction performance based on decision tree

6.5.5   *Performance Evaluation*

In this section, we evaluated the proposed framework and method based on a real EBSN dataset. Experiments were designed from two aspects, aiming to evaluate the weights of different features and the performance of different models.

6.5.5.1   Feature Evaluation

In the proposed multifactor model, user preference was characterized from three perspectives: content, spatiotemporal context, and social influence. To evaluate their influence to activity attendance prediction, we adopted the decision tree model [189] to perform attendance prediction based on each single category of features, and the results are shown in Fig. 6.28.

Accordingly, we find that models based on different features obtained distinct prediction performances. Specifically, the model based on content preference achieved the best performance with an accuracy of 0.79, while the model based on social influence was in the second place with an accuracy of 0.76.

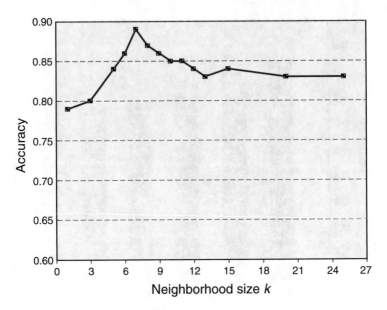

Fig. 6.29 Performance under different neighborhood size k

6.5.5.2   Model Evaluation

First, in the proposed SVD-MFN model, the main parameter that needs to be learned is the number of neighbors, which has a significant impact on the prediction performance [190]. To this end, we designed an experiment by varying the neighborhood size from 1 to 25; the corresponding results are shown in Fig. 6.29.

Accordingly, the size of the neighborhood does have a significant impact on the model's performance. Specifically, the prediction accuracy increased along with the increase of neighborhood size at the beginning, and then started to decrease. The reason might be that too many neighbors will result in too much noise. In our experiment, the highest prediction accuracy (0.89) was achieved when the neighborhood size was 7.

Afterwards, to examine the performance of SVD-MFN on attendance prediction, we compared it with three baseline methods, including decision tree, SVD, and ZeroR [191]. According to the experimental results shown in Fig. 6.30, we can see that SVD-MFN outperformed all the baseline methods, which proved the effectiveness of the proposed multifactor model. For more details about the experimental results, please refer to the article [192].

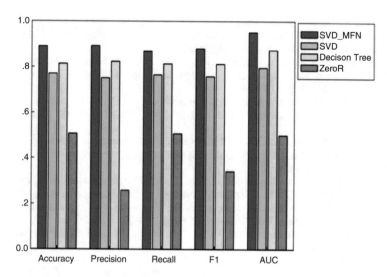

Fig. 6.30 Performance comparison with other approaches

References

1. R. Sen, Y. Lee, K. Jayarajah, A. Misra, and R. K. Balan, "GruMon: Fast and accurate group monitoring for heterogeneous urban spaces," in Proc. 12th ACM Conf. Embedded Netw. Sensor Syst., 2014, pp. 46–60.
2. N. Yu and Q. Han, "Grace: Recognition of proximity-based intentional groups using collaborative mobile devices," in Proc. 11th Int. Conf. Mobile Ad Hoc Sensor Syst., 2014, pp. 10–18.
3. Q. Li, Q. Han, X. Cheng, and L. Sun, "QueueSense: Collaborative recognition of queuing on mobile phones," in Proc. 11th Annu. IEEE Int. Conf. Sens. Commun. Netw., 2014, pp. 230–238.
4. M. B. Kjærgaard, M. Wirz, D. Roggen, and G. Troster, "Detecting pedestrian flocks by fusion of multi-modal sensors in mobile phones," in Proc. ACM Conf. Ubiquitous Comput., 2012, pp. 240–249.
5. M. B. Kjærgaard, M. Wirz, D. Roggen, and G. Troster, "Mobile sensing of pedestrian flocks in indoor environments using WiFi signals," in Proc. IEEE Int. Conf. Pervasive Comput. Commun., 2012, pp. 95–102.
6. M. Wirz, P. Schlapfer, M. B. Kjærgaard, D. Roggen, S. Feese, and G. Troster, "Towards an online detection of pedestrian flocks in urban canyons by smoothed spatio-temporal clustering of GPS trajectories," in Proc. 3rd ACM SIGSPATIAL Int. Workshop Location-Based Social Netw., 2011, pp. 17–24.
7. M. Costa, "Interpersonal distances in group walking," J. Nonverbal Behavior, vol. 34, no. 1, pp. 15–26, 2010.
8. M. Moussaid, N. Perozo, S. Garnier, D. Helbing, and G. Theraulaz, "The walking behaviour of pedestrian social groups and its impact on crowd dynamics," PloS One, vol. 5, no. 4, 2010, Art. no. e10047.
9. M. B. Kjærgaard, et al., "Time-lag method for detecting following and leadership behavior of pedestrians from mobile sensing data," in Proc. IEEE Int. Conf. Pervasive Comput. Commun., 2013, pp. 56–64.

10. B. Guo, Z. Yu, L. Chen, X. Zhou, and X. Ma, "MobiGroup: Enabling lifecycle support to social activity organization and suggestion with mobile crowd sensing,". IEEE Trans. Human-Mach Syst., vol. 46, no. 3, pp. 390–402, Jun. 2016.

11. S. Bandini, L. Crociani, A. Gorrini, and G. Vizzari, "An agent-based model of pedestrian dynamics considering groups: A real world case study," in Proc. 17th Int. IEEE Conf. Intell. Transp. Syst., 2014, pp. 572–577.

12. S. Bandini, L. Manenti, and S. Manzoni, "Generation of pedestrian groups distributions with probabilistic cellular automata," in Cellular Automata, Berlin, Germany: Springer, 2012, pp. 299–308.

13. S. Jamil, A. Basalamah, A. Lbath, and M. Youssef, "Hybrid participatory sensing for analyzing group dynamics in the largest annual religious gathering," in Proc. ACM Int. Joint Conf. Pervasive Ubiquitous Comput., 2015, pp. 547–558.

14. K. Jayarajah, Y. Lee, A. Misra, and R. K. Balan, "Need accurate user behaviour? pay attention to groups," in Proc. ACM Int. Joint Conf. Pervasive Ubiquitous Comput., 2015, pp. 855–866.

15. L. Chen, J. Hoey, C. Nugent, D. Cook, and Z. Yu, "Sensor-based activity recognition," IEEE Trans. Syst. Man Cybern. Part C, vol. 46, no. 6, pp. 790–808, Nov. 2012.

16. M. Azizyan, I. Constandache, and R. Roy Choudhury, "SurroundSense: Mobile phone localization via ambience fingerprinting," in Proc. 15th Annu. Int. Conf. Mobile Comput. Netw., 2009, pp. 261–272.

17. Z. Yu, H. Wang, B. Guo, T. Gu, and T. Mei, "Supporting serendipitous social interaction using human mobility prediction," IEEE Trans. Human-Mach Syst., vol. 45, no. 6, pp. 811–818, Dec. 2015.

18. H. Chen, B. Guo, Z. Yu, L. Chen, and X. Ma, "A generic framework for constraint-driven data selection in mobile crowd photographing," IEEE Internet Things J., vol. 4, no. 1, pp. 284–296, Feb. 2017.

19. Z. Yu, H. Xu, Z. Yang, and B. Guo, "Personalized travel package with multi-point-of-interest recommendation based on crowdsourced user footprints,". IEEE Trans. Human-Mach Syst., vol. 46, no. 1, pp. 151–158, Feb. 2016.

20. S. Feese, B. Arnrich, G. Troster, M. Burtscher, B. Meyer, and K. Jonas, "CoenoFire: Monitoring performance indicators of firefighters in real-world missions using smartphones," in Proc. ACM Int. Joint Conf. Pervasive Ubiquitous Comput., 2013, pp. 83–92.

21. S. Feese, B. Arnrich, G. Troster, M. Burtscher, B. Meyer, and K. Jonas, "Sensing group proximity dynamics of firefighting teams using smartphones," in Proc. Int. Symp. Wearable Comput., 2013, pp. 97–104.

22. D. Sanchez-Cortes, O. Aran, M. S. Mast, and D. Gatica-Perez, "A nonverbal behavior approach to identify emergent leaders in small groups," IEEE Trans. Multimedia, vol. 14, no. 3, pp. 816–832, Jun. 2012.

23. H. Du, Z. Yu, F. Yi, Z. Wang, Q. Han, and B. Guo, "Group mobility classification and structure recognition using mobile devices," in Proc. IEEE Conf. Pervasive Comput. Commun., 2016, pp. 1–9.

24. N. Yu, Y. Zhao, Q. Han, W. Zhu, and H. Wu, "Identification of Partitions in a Homogeneous Activity Group Using Mobile Devices," Mobile Information Systems, vol. 2016, Article ID 3545327, pp. 1–14.

25. Yu, Na, and Qi Han. "Context-aware community construction in proximity-based mobile networks." Mobile Information Systems 2015 (2015) 18.

26. Gordon, Dawud, et al. "Group affiliation detection using model divergence for wearable devices." Proceedings of the 2014 ACM International Symposium on Wearable Computers. ACM, 2014.

27. E. Dim and T. Kuflik, "Automatic detection of social behavior of museum visitor pairs," ACM Trans. Interactive Intell. Syst., vol. 4, no. 4, 2015, Art. no. 17.

28. L. Fosh, S. Benford, and B. Koleva, "Supporting group coherence in a museum visit," in Proc. 19th ACM Conf. Comput.-Supported Cooperative Work Social Comput., 2016, pp. 1–12.

29. A. Stisen, A. Mathisen, S. K. Sorensen, H. Blunck, M. B. Kjærgaard, and T. S. Prentow, "Task phase recognition for highly mobile workers in large building complexes," in Proc. IEEE Int. Conf. Pervasive Comput. Commun., 2016, pp. 1–9.

30. Martella, Claudio, et al. "Leveraging proximity sensing to mine the behavior of museum visitors." Pervasive Computing and Communications (PerCom), 2016 IEEE International Conference on. IEEE, 2016.

31. D. Wu, Y. Ke, J. X. Yu, S. Y. Philip, and L. Chen, "Detecting leaders from correlated time series," in Database Systems for Advanced Applications, Berlin, Germany: Springer, 2010, pp. 352–367.

32. Y. Sakurai, S. Papadimitriou, and C. Faloutsos, "BRAID: Stream mining through group lag correlations," in Proc. SIGMOD Int. Conf. Manage. Data, 2005, pp. 599–610.

33. Beyan, Cigdem, et al. "Multi-task learning of social psychology assessments and nonverbal features for automatic leadership identification." Proceedings of the 19th ACM International Conference on Multimodal Interaction. ACM, 2017.

34. B. Guo, H. Chen, Q. Han, Z. Yu, D. Zhang, and Y. Wang, "Worker-contributed data utility measurement for visual crowdsensing systems," IEEE Trans. Mobile Comput., to be published. Doi: https://doi.org/10.1109/TMC.2016.2620980.

35. B. Guo, Y. Liu, W. Wu, Z. Yu, and Q. Han, "ActiveCrowd: A framework for optimized multi-task allocation in mobile crowdsensing systems," IEEE Trans. Human-Mach. Syst., to be published. Doi: https://doi.org/10.1109/THMS.2016.2599489.

36. U. Blanke, G. Troster, T. Franke, and P. Lukowicz, "Capturing crowd dynamics at large scale events using participatory GPS localization," in Proc. IEEE 9th Int. Conf. Intell. Sensors Sensor Netw. Inf. Process., 2014, pp. 1–7.

37. M. Versichele, T. Neutens, M. Delafontaine, and N. Van de Weghe, "The use of Bluetooth for analysing spatiotemporal dynamics of human movement at mass events: A case study of the Ghent Festivities," Appl. Geography, vol. 32, no. 2, pp. 208–220, 2012.

38. M. Wirz, T. Franke, D. Roggen, E. Mitleton-Kelly, P. Lukowicz, and G. Troster, "Probing crowd density through smartphones in city-scale mass gatherings," EPJ Data Sci., vol. 2, no. 1, pp. 1–24, 2013.

39. A. Stisen, et al., "Smart devices are different: Assessing and mitigating mobile sensing heterogeneities for activity recognition," in Proc. 13th ACM Conf. Embedded Netw. Sensor Syst., 2015, pp. 127–140.

40. C. Luo and M. C. Chan, "SocialWeaver: Collaborative inference of human conversation networks using smartphones," in Proc. 11th ACM Conf. Embedded Netw. Sensor Syst., 2013, Art. no. 20.

41. Y. Lee, et al., "SocioPhone: Everyday face-to-face interaction monitoring platform using multi-phone sensor fusion," in Proc. 11th Annu. Int. Conf. Mobile Syst. Appl. Services, 2013, pp. 375–388.

42. M. R. Morris, K. Inkpen, and G. Venolia, "Remote shopping advice: Enhancing in-store shopping with social technologies," in Proc. 17th ACM Conf. Comput. Supported Cooperative Work Social Comput., 2014, pp. 662–673.

43. R. H. Shumway and D. S. Stoffer, Time Series Analysis and Its Applications with R Examples. Berlin, Germany: Springer, 2010.

44. F. Li, C. Zhao, G. Ding, J. Gong, C. Liu, and F. Zhao, "A reliable and accurate indoor localization method using phone inertial sensors," in Proc. ACM Conf. Ubiquitous Comput., 2012, pp. 421–430.

45. H. Lu, W. Pan, N. D. Lane, T. Choudhury, and A. T. Campbell, "SoundSense: Scalable sound sensing for people-centric applications on mobile phones," in Proc. 7th Int. Conf. Mobile Syst. Appl. Services, 2009, pp. 165–178.

46. W. Gu, Z. Yang, L. Shangguan, W. Sun, K. Jin, and Y. Liu, "Intelligent sleep stage mining service with smartphones," in Proc. ACM Int. Joint Conf. Pervasive Ubiquitous Comput., 2014, pp. 649–660.

47. H. Du, Z. Yu, F. Yi, Z. Wang, Q. Han and B. Guo, "Recognition of Group Mobility Level and Group Structure with Mobile Devices," in IEEE Transactions on Mobile Computing, vol. 17, no. 4, pp. 884–897, 1 April 2018. DOI: https://doi.org/10.1109/TMC.2017.2694839

48. M. Doyle and D. Straus, How to Make Meetings Work, Berkley Publishing Group, 1993.

49. C. Nass and S. Brave Voice Activated: How People Are Wired for Speech and How Computers Will Speak with Us, MIT Press, 2004.

50. P. Henline, "Eight Collaboratory Summaries," Interactions, vol. 5, no. 3, 1998, pp. 66–72.

51. Chang, Xiaojun, et al. "Bi-level semantic representation analysis for multimedia event detection." IEEE transactions on cybernetics 47.5 (2017): 1180–1197.

52. Zhu, Lei, et al. "Unsupervised visual hashing with semantic assistant for content-based image retrieval." IEEE Transactions on Knowledge and Data Engineering 29.2 (2017): 472–486.

53. Shah, Rajiv Ratn. "Multimodal analysis of user-generated content in support of social media applications." Proceedings of the 2016 ACM on International Conference on Multimedia Retrieval. ACM, 2016.

54. Wu, Zuxuan, et al. "Harnessing object and scene semantics for large-scale video understanding." Proceedings of the IEEE Conference on Computer Vision and Pattern Recognition. 2016.

55. Heilbron, Fabian Caba, et al. "SCC: Semantic Context Cascade for Efficient Action Detection." CVPR. 2017.

56. Z.W. Yu et al., "Capture, Recognition, and Visualization of Human Semantic Interactions in Meetings," Proc. Eighth Ann. IEEE Int'l Conf. Pervasive Computing and Communications (PerCom 10), 2010, pp. 107–115.

57. Z.W. Yu and Y. Nakamura, "Smart Meeting Systems: A Survey of State-of-the-Art and Open Issues," ACM Computing Surveys, vol. 42, no. 2, 2010, article 8.

58. Z.W. Yu et al., "Tree-Based Mining for Discovering Patterns of Human Interaction in Meetings," IEEE Trans. Knowledge and Data Eng., vol. 24, no. 4, 2012, pp. 759–768.

59. J. R. Hackman, Groups That Work (and Those That Don't), Jossey-Bass, 1990.

60. F. Pianesi et al., "Multimodal Support to Group Dynamics," Personal and Ubiquitous Computing, vol. 12, no. 3, 2008, pp. 181–195.

61. Dey, A K., Salber, D., Abowd, G. D., and Futakawa, M. The Conference Assistant: Combining Context-Awareness with Wearable Computing, In Proc. ISWC'99, 21–28.

62. Sumi, Y. and Mase, K. Digital Assistant for Supporting Conference Participants: An Attempt to Combine Mobile, Ubiquitous and Web Computing. In Proc. Ubicomp 2001, 156–175.

63. Chen, H., Finin, T., and Joshi, A A Context Broker for Building Smart Meeting Rooms. In Proc. of the Knowledge Representation and Ontology for autonomous systems symposium (AAAI spring symposium), AAAI, 2004, 53–60.

64. Koike, H., Nagashima, S., Nakanishi, Y., and Sato, Y.: Enhanced Table Supporting a Small Meeting in Ubiquitous and Augmented Environment, In Proc. PCM 2004, 97–104.

65. Chiu, P., Kapuskar, A, Reitmeier, S., and Wilcox, L. Room with a Rear View: Meeting Capture in a Multimedia Conference Room, IEEE Multimedia, 7(4), 2000, 48–54.

66. Richter, H., Abowd, G. D., Geyer W., Fuchs, L., Daijavad, S., and Poltrock, S. Integrating Meeting Capture within a Collaborative Team Environment. In Proc. Ubicomp 2001, 123–138.

67. Ark, W. S. and Selker, T. A look at human interaction with pervasive computers. IBM Systems Journal, 38(4), 1999, 504–507

68. Vapnik, V. N. The Nature of Statistical Learning Theory. Springer Verlag, Heidelberg, DE, 1995.

69. Ahmed, S., Sharmin, M., and Ahmed, S. I. A Smart Meeting Room with Pervasive Computing Technologies. In Proc. SNPDISAWN'05, IEEE Computer Society Press (2005),366–371.

70. Mikic, I., Huang, K., and Trivedi, M. Activity Monitoring and Summarization for an Intelligent Meeting Room. IEEE Workshop on Human Motion, 2000, 107–112.

71. Kim, N., Han, S., and Kim, J. W. Design of Software Architecture for Smart Meeting Space. In Proc. PerCom 2008, IEEE Press (2008), 543–547.

72. Araki, Shoko, et al. "Spatial correlation model based observation vector clustering and MVDR beamforming for meeting recognition." Acoustics, Speech and Signal Processing (ICASSP), 2016 IEEE International Conference on. IEEE, 2016.

73. Araki, Shoko, et al. "Online meeting recognition in noisy environments with time-frequency mask based MVDR beamforming." Hands-free Speech Communications and Microphone Arrays (HSCMA), 2017. IEEE, 2017.

74. Yun, Kiwon, et al. "Two-person interaction detection using body-pose features and multiple instance learning." Computer Vision and Pattern Recognition Workshops (CVPRW), 2012 IEEE Computer Society Conference on. IEEE, 2012.

75. Ji, Yanli, Guo Ye, and Hong Cheng. "Interactive body part contrast mining for human interaction recognition." Multimedia and Expo Workshops (ICMEW), 2014 IEEE International Conference on. IEEE, 2014.

76. Zhu, Wentao, et al. "Co-Occurrence Feature Learning for Skeleton Based Action Recognition Using Regularized Deep LSTM Networks." AAAI. Vol. 2. No. 5. 2016.

77. Stiefelhagen, R., Chen, X, and Yang, J. Capturing Interactions in Meetings with Omnidirectional Cameras. International Journal of Distance Education Technologies, 3(3), 2005, 34–47.

78. Nijholt, A, Rienks, R J., Zwiers, J., and Reidsma, D. Online and Off-line Visualization of Meeting Information and Meeting Support. The Visual Computer, 22 (12), 2006, 965–976.

79. Kim, T., Chang, A, Holland, L., and Pentland, A Meeting Mediator: Enhancing Group Collaboration using Sociometric Feedback. In Proc. CSCW 2008, 457–466.

80. Sumi, Y., et al. Collaborative capturing, interpreting, and sharing of experiences, Personal and Ubiquitous Computing, 11(4), 2007, 265–271.

81. DiMicco, J. M., et al. The Impact of Increased Awareness While Face to-Face. Human-Computer Interaction, 22(1),47–96 (2007)

82. Otsuka, K., Sawada, H., and Yamato, J. Automatic Inference of Cross modal Nonverbal Interactions in Multiparty Conversations. In Proc. ICMI 2007, 255–262.

83. Dielmann, A and Renals, S. Dynamic Bayesian Networks for Meeting Structuring. In Proc. 1CASSP2004, 629–632.

84. McCowan, I., Gatica-Perez, D., Bengio, S., Lathoud, G., Bamard, M., and Zhang, D. Automatic Analysis of Multimodal Group Actions in Meetings. IEEE Transactions on Pattern Analysis and Machine Intelligence, 27(3), 2005, 305–317.

85. Rybski, P. E. and Veloso, M. M. Using Sparse Visual Data to Model Human Activities in Meetings. In Proc. of IJCA1 Workshop on Modeling Other Agents from Observations (MOO 2004).

86. Hillard, D., Ostendorf, M., and Shriberg, E. Detection of Agreement vs. Disagreement in Meetings: Training with Unlabeled Data. In Proc. HLT NAACL 2003, 34–36.

87. Tomobe, H. and Nagao, K. Discussion Ontology : Knowledge Discovery from Human Activities in Meetings. In Proc. JSAI 2006, 33–41.

88. Garg, N. P., Favre, S., Salamin, H., Tur, D. H., and Vinciarelli, A Role Recognition for Meeting Participants: an Approach Based on Lexical Information and Social Network Analysis. In Proc. ACM Multimedia 2008, 693–696.

89. Laurent, Antoine, Nathalie Camelin, and Christian Raymond. "Boosting bonsai trees for efficient features combination: application to speaker role identification." Interspeech. 2014.

90. Searle, J. Speech Acts, Cambridge University Press, 1969.

91. Waibel, A, Bett, M., and Finke, M. Meeting Browser: Tracking and Summarizing Meetings. Proc. of the Broadcast News Transcription and Understanding Workshop, Lansdowne, Virginia, February 1998, 281–286.

92. Colbath, S. and Kubala, F. Rough'n'Ready: A Meeting Recorder and Browser. In Proc. of the Perceptual User Interface Conference, San Francisco, CA, November 4–6, 1998, 220–223.

93. Wellner, P., Flynn, M., and Guillemot, M. Browsing Recorded Meetings with Ferret. Proc. of the First International Workshop on Machine Learning for Multimodal Interaction (MLMI'04), Martigny, Switzerland, June 21–23, 2004, 12–21.

94. Geyer W., Richter, H., and Abowd, G. D. Towards a Smarter Meeting Record Capture and Access of Meetings Revisited. Multimedia Tools and Applications, 27(3), 2005, 393–410.

95. Jaimes, A, Omura, K., Nagamine, T., and Hirata, K. Memory Cues for Meeting Video Retrieval. The first ACM Workshop on Continuous Archival and Retrieval of Personal Experiences (CARPE'04), New York, NY, USA, October IS, 2004, 74–85.

96. Junuzovic, S., Hegde, R., Zhang, Z., Chou, P. A., Liu, Z., and Zhang, C. Requirements and Recommendations for an Enhanced Meeting Viewing Experience. In Proc. of ACM Multimedia 2008, 539–548.

97. Araki, M., Itoh, T., Kumagai, T., and Ishizaki, M. Proposal of a standard utterance unit tagging scheme. Journal of Japanese Society for artificial intelligence, 14(2), 1999, 251–260.

98. PhaseSpace IMPULSE system. http://www.phasespace.coml.

99. Julius speech recognition engine. http://julius.sourceforge.jp/enl.

100. Rabiner, L. A tutorial on Hidden Markov Models and selected applications in speech recognition. In Proc. IEEE, 77(2), 1989, 257–286.

101. Kipp, M. Anvil - A Generic Annotation Tool for Multimodal Dialogue. In Proc. Eurospeech 2001, 1367–1370.

102. Chang, C. C., and Lin, C. J. LIBSVM: a library for support vector machines, 2001. Software available at http://www.csie.ntu.edu.tw/-cjlinllibsvm

103. Yu, H., Finke, M., and Waibel, A. Progress in Automatic Meeting Transcription. Proc. of 6th European Conference on Speech Communication and Technology (Eurospeech-99), September 5–9, 1999, Budapest, Hungary, Vol. 2, 695–698.

104. Z. Yu, Z. Yu, H. Aoyama, M. Ozeki and Y. Nakamura, "Capture, recognition, and visualization of human semantic interactions in meetings," 2010 IEEE International Conference on Pervasive Computing and Communications (PerCom), Mannheim, 2010, pp. 107–115. DOI: https://doi.org/10.1109/PERCOM.2010.5466987

105. A. C. Ellis, S. J. Gibbs, and G. Rein, "Groupware: Some issues and experiences," Commun. ACM, vol. 31, no. 1, pp. 39–58, 1991.

106. D. MacLean, S. Hangal, S. K. Teh, M. S. Lam, and J. Heer, "Groups without tears: Mining social topologies from email," in Proc. 16th Int. Conf. Intell. User Interface, 2011, pp. 83–92.

107. M. Roth, A. Ben-David, D. Deutscher, G. Flysher, I. Horn, A. Leichtberg, and R. Merom, "Suggesting friends using the implicit social graph," in Proc. 16th ACM SIGKDD Int. Conf., 2010, pp. 233–242.

108. Saleema, J. Fogarty, and D. Weld, "Regroup: Interactive machine learning for on-demand group creation in social networks," in Proc. SIGCHI Conf., 2012, pp. 21–30.

109. M. Eslami, A. Aleyasen, R. Z. Moghaddam, and K. G. Karahalios, "Evaluation of automated friend grouping in online social networks," in Proc. SIGCHI Conf., 2014, pp. 2119–2124.

110. S. Whittaker, Q. Jones, B. Nardi, M. Creech, L. Terveen, E. Isaacs, and J. Hainsworth, "ContactMap: Organizing communication in a social desktop," ACM Trans. Comput. Hum. Int., vol. 11, no. 4, pp. 445–471, 2004.

111. E. Miluzzo, N. D. Lane, K. Fodor, R. Peterson, H. Lu, M. Musolesi, and A. T. Campbell, "Sensing meets mobile social networks: The design, implementation and evaluation of the CenceMe application," in Proc. 6th ACM Conf. Emb. Netw. Sens. Syst., 2008, pp. 337–350.

112. X. Bao and R. R. Choudhury, "MoVi: Mobile phone-based video highlights via collaborative sensing," in Proc. 8th ACM Conf. Emb. Netw. Sens. Sys., 2010, pp. 357–370.

113. E. G. Boix, A. L. Carreton, C. Scholliers, T. V. Cutsem, W. D. Meuter, and T. D'Hondt, "Flocks: Enabling dynamic group interactions in mobile social networking applications," in Proc. ACM Symp. Appl. Comput., 2011, pp. 425–432.

114. B. Guo, D. Zhang, Z. Wang, Z. Yu, and X. Zhou, "Opportunistic IoT: Exploring the harmonious Interaction between human and the Internet of Things," J. Netw. Comput. Appl., vol. 36, no. 6, pp. 1531–1539, 2013.

115. D. Zhang, B. Guo, and Z. Yu, "The emergence of social and community intelligence," Comput., vol. 44, no. 7, pp. 21–28, 2011.

116. B. Guo, Z. Wang, Z. Yu, Y. Wang, N. Y. Yen, R. Huang, and X. Zhou, "Mobile crowd sensing and computing: The review of an emerging human-powered sensing paradigm," ACM Comput. Surv., vol. 48, no. 1, pp. 1–31, 2015.

117. B. Guo, H. He, Z. Yu, D. Zhang, and X. Zhou, GroupMe: Supporting group formation with mobile sensing and social graph mining," in Proc. Mob. Ubiq. Syst., 2012, pp. 200–211.

118. A. Jameson and B. Smyth, "Recommendation to groups," in Adaptive Web. New York, NY, USA: Springer, 2007, pp. 596–627.

119. B. Guo, Z. Yu, D. Zhang, and X. Zhou, "Cross-community sensing and mining," IEEE Commu. Mag., vol. 52, no. 8, pp. 144–152, Aug. 2014.

120. H. Z. Kim and K. S. Eklundh, "Reviewing practices in collaborative writing," Comput. Supp. Cooperative Work, vol. 10, no. 2, pp. 247–259, 2001.

121. Z. Yang, W. Wu, K. Nahrstedt, G. Kurillo, and R. Bajcsy, "Enabling multi-party 3D tele-immersive environments with ViewCast," ACM Trans. Multi. Comput., vol. 6, no. 2, pp. 1–30, 2010.

122. B. Guo, et al. "FlierMeet: a mobile crowdsensing system for cross-space public information reposting, tagging, and sharing", IEEE Transactions on Mobile Computing 1 (2015): 1–1.

123. Liu, Yan, et al. "FooDNet: Toward an Optimized Food Delivery Network based on Spatial Crowdsourcing." IEEE Transactions on Mobile Computing (2018).

124. Wang, Qianru, et al. "CrowdWatch: dynamic sidewalk obstacle detection using mobile crowd sensing." IEEE Internet of Things Journal 4.6 (2017): 2159–2171.

125. Pinkerton, Sean, et al. ""Those sweet, sweet likes": Sharing physical activity over social network sites." Computers in Human Behavior 69 (2017): 128–135.

126. E. Cho, S. A. Myers, and J. Leskovec, "Friendship and mobility: User movement in location-based social networks," in Proc. 17th ACM SIGKDD Int. Conf., 2011, pp. 1082–1090.

127. J. Tang, T. Lou, and J. Kleinberg, "Inferring social ties across heterogeneous networks," in Proc. 5th ACM Int. Conf. Web Search Data Mining, 2012, pp. 743–752.

128. B. Guo, D. Zhang, D. Yang, Z. Yu, and X. Zhou, "Enhancing memory recall via an intelligent social contact management system," IEEE Trans. Human Mach. Syst., vol. 44, no. 1, pp. 78–91, 2014.

129. N. Cross and A. C. Cross, "Observations of teamwork and social processes in design," Des. Stud., vol. 16, no. 2, pp. 143–170, 1995.

130. B. A. Nardi, Context and Consciousness: Activity Theory and Human Computer Interaction. Cambridge, MA, USA: MIT Press, 1996.

131. D. Easley and J. Kleinberg, Networks, Crowds, and Markets. Cambridge, U.K.: Cambridge Univ. Press, 2010.

132. S. Wakamiya, R. Lee, and K. Sumiya, "Urban area characterization based on semantics of crowd activities in twitter," in Proc. GeoSpatial Semantics, 2011, pp. 108–123.

133. Q. Chuan, X. Bao, R. R. Choudhury, and S. Nelakuditi, "Tagsense: A smartphone-based approach to automatic image tagging," in Proc. 9th Int. Conf. Mobile Syst., Appl. Serv., 2011, pp. 1–14.

134. H. V. Enrique, S. Alonso, F. Chiclana, and F. Herrera, "A consensus model for group decision making with incomplete fuzzy preference relations," IEEE Trans. Fuzzy Syst., vol. 15, no. 5, pp. 863–877, Oct. 2007.

135. M. E. J. Newman, "Modularity and community structure in networks," Proc. Natl. Acad. Sci., vol. 103, no. 23, pp. 8577–8582, 2006.

136. J. Staiano, B. Lepri, N. Aharony, F. Pianesi, N. Sebe, and A. Pentland, "Friends don't lie: Inferring personality traits from social network structure," in Proc. ACM Conf. Ubi. Comput., 2012, pp. 321–330.

137. A. K. Dey, "Understanding and using context," Pers. Ubiquit. Comput., vol. 5, no. 1, pp. 4–7, 2001.

138. C. Bizer, J. Lehmann, G. Kobilarov, S. Auer, C. Becker, R. Cyganiak, and S. Hellmann, "DBpedia-A crystallization point for the web of data," J. Web Semantics, vol. 7, no. 3, pp. 154–165, 2009.

139. Q. Yuan, G. Cong, Z. Ma, A. Sun, and N. M. Thalmann, "Time-aware point-of-interest recommendation," in Proc. 36th Int. ACM SIGIR Conf., 2013, pp. 363–372.

140. B. Guo, X. Xie, H. Chen, S. Huangfu, Z. Yu, and Z. Wang, "FlierMeet: Cross-space public information reposting with mobile crowd sensing," in Proc. ACM Conf. Ubi. Comp. Adjunc., 2014, pp. 59–62.

141. K. Cheverst, K. Mitchell, and N. Davies, "Investigating context-aware information push vs. information pull to tourists," in Proc. Mobile HCI, 2001, pp. 1–6.

142. J. Yang and J. Leskovec, "Patterns of temporal variation in online media," in Proc. 4th ACM Int. Conf. Web Search Data Mining, 2011, pp. 177–186.

143. R. Kumar, J. Novak, and A. Tomkins, "Structure and evolution of online social networks," Link Mining: Models, Algorithms, Applications. New York, NY, USA: Springer, 2010, pp. 337–357.

144. Q. Yang, S. J. Pan, and V. W. Zheng, "Estimating location using Wi-Fi," IEEE Intell. Syst., vol. 23, no. 1, pp. 8–13, Jan./Feb. 2008.

145. E. Tapia, S. S. Intille, and K. Larson, "Activity recognition in the home using simple and ubiquitous sensors," in Proc. Int. Conf. Pervasive Comput., 2004, pp. 158–175.

146. N. Ihaddadene, M. H. Sharif, and C. Djeraba, "Crowd behaviour monitoring," in Proc. 16th ACM Int. Conf. Multimedia, 2008, pp. 1013–1014.

147. X. Bao and R. R. Choudhury, "VUPoints: Collaborative sensing and video recording through mobile phones," ACM SIGCOMM Comput. Commun. Review, vol. 40, no. 1, pp. 100–105, 2010.

148. M. Azizyan, I. Constandache, and R. R. Choudhury, "SurroundSense: Mobile phone localization via ambience fingerprinting," in Proc. 15th ACM Int. Conf. Mobile Comput. Netw., 2009, pp. 261–272.

149. R. Jang. (Jun. 2008). Audio signal processing and recognition. [Online]. Available: http://mirlab.org/jang/books/audioSignalProcessing

150. T. Zhang and C. Kuo, "Hierarchical classification of audio data for archiving and retrieving," in Proc. Int. Conf. Acoust., Speech, Signal Process., 1999, pp. 3001–3004.

151. G. Tzanetakis and P. Cook, "Sound analysis using MPEG compressed audio," in Proc. Int. Conf. Acoust. Speech, Signal Process., 2000, pp. 11761–11764.

152. Y. Zhan and T. Kuroda, "Wearable sensor-based human activity recognition from environmental background sounds," J. Amb. Intel. Human Comput., vol. 5, no. 1, pp. 77–89, 2014.

153. D. J. Berndt and J. Clifford, "Using dynamic time warping to find patterns in time series," in Proc. ACM KDD Workshops, 1994, vol. 10, no. 16, pp. 359–370.

154. M. Mitchell, "Complex systems: Network thinking," Artif. Intell., vol. 170, no. 18, pp. 1194–1212, 2006.

155. B. Guo, Z. Yu, L. Chen, X. Zhou and X. Ma, "MobiGroup: Enabling Lifecycle Support to Social Activity Organization and Suggestion With Mobile Crowd Sensing," in IEEE Transactions on Human-Machine Systems, vol. 46, no. 3, pp. 390–402, June 2016. DOI: https://doi.org/10.1109/THMS.2015.2503290

156. X. Liu, Q. He, Y. Tian, W.-C. Lee, J. McPherson, and J. Han. Event-based social networks: linking the online and offline social worlds. In Knowledge Discovery and Data Mining, pages 1032–1040, 2012.

157. Pramanik, S., Gundapuneni, M., Pathak, S., & Mitra, B. (2016, March). Predicting Group Success in Meetup. In ICWSM (pp. 663–666).

158. J. Han, J. Niu, A. Chin, W. Wang, C. Tong, and X. Wang. How online social network affects offline events: A case study on Douban. In Ubiquitous Intelligence & Computing and International Conference on Autonomic & Trusted Computing, pages 752–757, 2012.

159. B. Xu, A. Chin, and D. Cosley. On how event size and interactivity affect social networks. In CHI Extended Abstracts on Human Factors in Computing Systems, pages 865–870, 2013.

160. D. H. Lee. PITTCULT: trust-based cultural event recommender. In Conference on Recommender Systems, pages 311–314, 2008.

161. Cao, Jiuxin, et al. "Multi-feature based event recommendation in Event-Based Social Network." Int. J. Comput. Intell. Syst. 11.1 (2018): 618–633.
162. Liu, Shenghao, Bang Wang, and Minghua Xu. "Event recommendation based on graph random walking and history preference reranking." Proceedings of the 40th international ACM SIGIR conference on research and development in information retrieval. ACM, 2017.
163. Bixi Li, Bang Wang, Yijun Mo and Laurence T. Yang. A Novel Random Walk and Scale Control Method for Event Recommendation. The 2016 IEEE Smart World Congress, pp. 228–235, 2016.
164. R. Klamma, P. M. Cuong, and Y. Cao. You never walk alone: Recommending academic events based on social network analysis. In Complex Sciences, pages 657–670. 2009.
165. C. Cornelis, X. Guo, J. Lu, and G. Zhang. A Fuzzy Relational Approach to Event Recommendation. In Indian International Conference on Artificial Intelligence, pages 2231–2242, 2005.
166. E. M. Daly and W. Geyer. Effective event discovery: using location and social information for scoping event recommendations. In Proceedings of ACM Conference on Recommender Systems, pages 277–280. ACM, 2011.
167. E. Minkov, B. Charrow, J. Ledlie, S. J. Teller, and T. Jaakkola. Collaborative future event recommendation. In International Conference on Information and Knowledge Management, pages 819–828, 2010.
168. J. Sang, T. Mei, J.-T. Sun, C. Xu, and S. Li. Probabilistic sequential POIs recommendation via check-in data. In Proceedings of International Conference on Advances in Geographic Information Systems, pages 402–405, 2012.
169. J. Zhuang, T. Mei, S. C. Hoi, Y.-Q. Xu, and S. Li. When recommendation meets mobile: contextual and personalized recommendation on the go. In Proceedings of the ACM International Conference on Ubiquitous Computing, pages 153–162, 2011.
170. Z. Wang, D. Zhang, X. Zhou, D. Yang, Z. Yu, and Z. Yu. Discovering and profiling overlapping communities in location-based social networks. Systems, Man, and Cybernetics: Systems, IEEE Transactions on, 44(4):499–509, April 2014.
171. Z. Yu, Y. Yang, X. Zhou, Y. Zheng, and X. Xing. Investigating how user's activities in both virtual and physical world impact each other leveraging LBSN data. International Journal of Distributed Sensor Networks, vol. 2014, Article ID 461780, pp. 1–9, 2014.
172. E. Cho, S. A. Myers, and J. Leskovec. Friendship and mobility: user movement in location-based social networks. In Knowledge Discovery and Data Mining, pages 1082–1090. ACM, 2011.
173. A. Noulas, S. Scellato, C. Mascolo, and M. Pontil. An empirical study of geographic user activity patterns in foursquare. International Conference on Weblogs and Social Media, pages 70–573, 2011.
174. D. J. Crandall, L. Backstrom, D. Cosley, S. Suri, D. Huttenlocher, and J. Kleinberg. Inferring social ties from geographic coincidences. Proceedings of the National Academy of Sciences, 107(52):22436–22441, 2010.
175. H. Zhuang, A. Chin, S. Wu, W. Wang, X. Wang, and J. Tang. Inferring geographic coincidence in ephemeral social networks. In Machine Learning and Knowledge Discovery in Databases, pages 613–628. 2012.
176. Yi, Fei, et al. "An Opportunistic Music Sharing System Based on Mobility Prediction and Preference Learning." 2014 IEEE 11th Intl Conf on Ubiquitous Intelligence & Computing UIC 2014.
177. D. Bollegala, Y. Matsuo, and M. Ishizuka. Measuring semantic similarity between words using web search engines. In World Wide Web Conference Series, pages 757–766, 2007.
178. Y. Li, D. Mclean, Z. A. Bandar, JD O'Shea, and K. A. Crockett. Sentence Similarity Based on Semantic Nets and Corpus Statistics. IEEE Transactions on Knowledge and Data Engineering, 18:1138–1150, 2006.
179. D. M. Blei, A. Y. Ng, and M. I. Jordan. Latent Dirichlet allocation. Advances in neural information processing systems, 1:601–608, 2002.

180. J. Lin. Divergence measures based on the Shannon entropy. IEEE Transactions on Information Theory, 37:145–151, 1991.
181. T. L. Griffiths. Finding scientific topics. Proceedings of The National Academy of Sciences, 101:5228–5235, 2004.
182. Y. Cheng, G. Qiu, J. Bu, K. Liu, Y. Han, C. Wang, and C. Chen. Model bloggers' interests based on forgetting mechanism. In World Wide Web Conference Series, pages 1129–1130, 2008.
183. D. Yang, D. Zhang, Z. Yu, and Z. Yu. Fine-grained preference-aware location search leveraging crowdsourced digital footprints from LBSNs. In Proceedings of the 2013 ACM international joint conference on Pervasive and ubiquitous computing, pages 479–488, 2013.
184. M. Ye, X. Liu, and W.-C. Lee. Exploring social influence for recommendation: a generative model approach. In Proceedings of ACM International Conference on Research and Development in Information Retrieval, pages 671–680, 2012.
185. L. Backstrom, D. P. Huttenlocher, J. M. Kleinberg, and X. Lan. Group formation in large social networks: membership, growth, and evolution. In Knowledge Discovery and Data Mining, pages 44–54, 2006.
186. Y. Koren, R. Bell, and C. Volinsky. Matrix factorization techniques for recommender systems. Computer, 42(8):30–37, 2009.
187. A. K. Menon, K.-P. Chitrapura, S. Garg, D. Agarwal, and N. Kota. Response prediction using collaborative filtering with hierarchies and side-information. In Knowledge Discovery and Data Mining, pages 1032–1040, 2012.
188. Y.-M. Li, C.-T. Wu, and C.-Y. Lai. A social recommender mechanism for e-commerce: Combining similarity, trust, and relationship. Decision Support Systems, 55(3):740–752, 2013.
189. M. Hall, E. Frank, G. Holmes, B. Pfahringer, P. Reutemann, and I. H. Witten. The WEKA data mining software: an update. SIGKDD Explorations, 11:10–18, 2009.
190. J. L. Herlocker, J. A. Konstan, A. Borchers, and J. Riedl. An algorithmic framework for performing collaborative filtering. In Research and Development in Information Retrieval, pages 230–237, 1999.
191. A. Ben-David. Comparison of classification accuracy using Cohen's weighted kappa. Expert Systems with Applications, 34(2):825–832, 2008.
192. R. Du, Z. Yu, T. Mei, Z.T. Wang, Z. Wang, B Guo Predicting activity attendance in event-based social networks: content, context and social influence, Proceedings of the 2014 ACM International Joint Conference on Pervasive and Ubiquitous Computing, Seattle, USA, September, 2014, pp. 425–434.

Chapter 7
Community Behavior Understanding

Abstract The recent rapid development of smart mobile devices and mobile social networking services makes it possible to explore human behaviors in an unprecedented large scale. In this chapter, we present some of our recent research advances on community behavior understanding. Specifically, in Sect. 7.1, we present the discovering and profiling communities in mobile social networks, followed by a study on how to understand the evolution of social relationships in Sect. 7.2. Finally, in Sect. 7.3, we discuss how to enhance human social interactions by interlinking off-line and online communities.

7.1 Discovering Communities in Mobile Social Networks[1]

7.1.1 Introduction

With the wide adoption of GPS-enabled smartphones, location-based social networks (LBSNs) have been experiencing increasing popularity, attracting millions of users. In LBSNs, users can explore places, write reviews, upload photos, and share locations and experiences with others. The soaring popularity of LBSNs has created opportunities for understanding collective user behaviors on a large scale, which are capable of enabling many applications, such as direct marketing, trend analysis, group search, and tracking.

One fundamental issue in social network analysis is to detect user communities. A community is typically thought of as a group of users with more and/or better interactions among its members than between its members and the remainder of the network [1, 2]. However, unlike social networks (e.g., Flickr, Facebook) which provide explicit groups for users to subscribe or join, the notion of community in

---

[1]Part of this section is based on a previous work: Z. Wang, D. Zhang, X. Zhou, D. Yang, Z. Yu and Z. Yu, "Discovering and Profiling Overlapping Communities in Location-Based Social Networks," in IEEE Transactions on Systems, Man, and Cybernetics: Systems, vol. 44, no. 4, pp. 499–509, April 2014. DOI: https://doi.org/10.1109/TSMC.2013.2256890.

LBSNs is not well defined. In order to capitalize on the huge number of potential users, quality community detection and profiling approaches are needed.

It has been well understood that people in a real social network are naturally characterized by multiple community memberships. For example, a person usually belongs to several social groups like family, friends, and colleges; a researcher may be active in several areas. Thus, it is more reasonable to cluster users into overlapping communities rather than disjoint ones.

Most of the existing community detection approaches are based on structural features (e.g., links) [3], but the structural information of online social networks is often sparse and weak, and thus it is difficult to detect interpretable overlapping communities by considering only network structural information [4]. Fortunately, LBSNs provide rich information about the user and venue through check-ins, which makes it possible to cluster users with different preferences and interests into different communities. Specifically, the observation that a check-in on LBSNs reflects a certain aspect of the user's preferences or interests enlightens us to cluster edges instead of nodes, as the detected clusters of check-ins will naturally assign users into overlapping communities with connections to venues. Once edge clusters are obtained, overlapping communities of users can be recovered by replacing each edge with its vertices; that is, a user is involved in a community as long as any of his/her check-ins falls into the community. In such a way, the obtained communities are usually highly overlapped.

We present an example of the user-venue check-in network in Fig. 7.1, which consists of five users and four venues. In such a network, users and venues are represented as two types of nodes, and each check-in is represented as an edge between a user node and a venue node. For this attributed bipartite network, since both users and venues have their own attributes, if we perform edge clustering to group users based solely on network structure [5], we can get two overlapping communities: Group 1 (Mary, Tom) and Group 2 (Tom, David, Bob, Eva). By

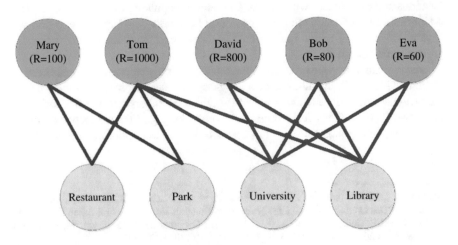

Fig. 7.1 A user-venue check-in network example

implicitly using the venue mode to characterize the user mode (i.e., inter-mode), we can interpret Group 1 as a family community and Group 2 as a colleague community. However, if we consider not only the check-in network (i.e., inter-mode features) but also the attributes of users and venues (i.e., intra-mode features), we can get three overlapping communities: Group 1 (Mary, Tom), Group 2 (Tom, David), and Group 3 (Bob, Eva). In this case, even though Tom, David, Bob, and Eva have similar check-in patterns, they are further grouped into two separate communities. Since Tom and David travel frequently the radius of gyration (i.e., r_g) are 1000 and 800 km, while Bob and Eva mainly stay locally whose r_g are 80 and 60 km, respectively. Here we probably can label Group 1 as a family community, Group 2 as a research staff community, and Group 3 as a teaching staff community.

Apparently it is more reasonable to exploit both the structural information (inter-mode) and the node attributes (intra-mode) to cluster users, as we can naturally obtain communities with richer and interpretable information, even though it is a highly challenging task. While classical co-clustering is one way to conduct this kind of community partitioning [6], the identified communities are disjoint which contradicts with the actual social setting. Edge clustering has been proposed to detect communities in an overlapping manner [5], but it did not take intra-mode features into consideration.

The objective of community detection in LBSNs is to discover users who share similar interests and at the same time are structurally close to each other. Therefore, the existing evaluation metrics (e.g., maximum modularity [1, 7] and normalized mutual information [5]) for community detection, which are designed from the network topology perspective, are not appropriate for evaluating the quality of the detected communities in LBSNs. Furthermore, existing evaluation metrics are not able to characterize the member's participation degree within a community. Thereby, novel evaluation metrics and mechanisms are needed, which poses another research challenge. In summary, the main contributions of this work are as follows:

- We formulate the overlapping community detection problem in LBSNs as a co-clustering issue which considers both the user-venue check-in network structure and attributes of users and venues. To the best of our knowledge, this work is the first attempt addressing the overlapping community detection problem in LBSNs. Specifically, we detect overlapping communities from an edge-centric perspective, with the attributes of users and venues considered.
- We represent users and venues in LBSNs as two types of modes (nodes), and select both inter-mode and intra-mode features for co-clustering, while existing multimode clustering methods mainly concern the inter-mode features. We show that different perspectives of social communities can be revealed by introducing different intra-mode features.
- We propose a novel community structure evaluation framework which consists of two metrics to ensure optimal community detection. While community interest entropy measures the similarity of community members, community participation entropy estimates how equal users engage in a community. Meanwhile, this framework can also be used to guide the selection of k (number of communities) for community detection.

7.1.2   Related Work

In this section, we briefly review the related work which can be classified into three categories. The first category contains the research on understanding the collective user behaviors based on Foursquare and other LBSN dataset. Scellato et al. [8, 9] analyzed the social, geographic, and geo-social properties of four social networks (BrightKite, Foursquare, LiveJournal, and Twitter). Noulas et al. [10] investigated the user check-in dynamics and the presence of spatiotemporal patterns in Foursquare. Similarly, Chen et al. [11] measured the similarities between users by constructing models to extract users' spatial and temporal features. Cheng et al. [12] studied the mobility patterns of Foursquare users and revealed the factors affecting people's mobility. He et al. [13] proposed a method to recommend the next point of interest (POI) based on Foursquare. Vasconcelos et al. [14] analyzed how Foursquare users exploited three features (i.e., tips, dones and to-dos) to uncover different behavior profiles.

Only two studies aimed at uncovering group profiles in LBSNs. Li et al. [15] proposed two different clustering approaches to identify user behavior patterns on BrightKite. One approach exploited the update (i.e., check-ins, photos, and notes) of users to classify them into four disjoint groups according to their mobility, namely home, home-vacation, home-work, and others. The second approach clustered users based on attributes such as total number of updates, social features, and mobility characteristics, and led to the identification of five disjoint groups, namely inactive, normal, active, mobile, and trial users. The second study was performed on Foursquare. Noulas et al. [16] used a spectral clustering algorithm to group users based on the categories of venues they had checked in, aiming at identifying communities and characterizing the type of activity in each region of a city. Although the aforementioned studies offer important insights into properties of user interactions in LBSNs, none of them worked on overlapping community detection using network links and node attributes. Our work aims to fill in this gap by discovering and profiling these overlapping communities, so as to enable direct and group-oriented marketing businesses.

The second category involves the work on community detection which is a classical task in complex network analysis [1, 2, 17, 18]. A community is typically thought of as a group of nodes with more and/or better interactions among its members than between its members and the remainder of the network [1, 2]. To extract such sets of nodes, one typically chooses an objective function that captures the intuition of a network cluster as set of nodes with better internal connectivity than external connectivity, and then one applies approximation or heuristics algorithms to extract node clusters by optimizing the objective function and revealing good communities for the application of interest. In addition, some researchers conduct community detection by combining embedding concept. For instance, Cavallari et al. [19] studied a closed loop among community embedding, community detection, and node embedding. In general, community detection can be classified into two types: overlapping and nonoverlapping methods. Some popular methods are modularity maximization [17, 18], Girvan-Newman algorithm [1], Louvain algorithm [20], clique percolation [21], link communities [3], etc. As users in LBSNs have rather weak and sparse relations [22], one cannot naively apply community detection based solely on the links found in these social networks and expect to generate interpretable communities.

The third category focuses on community detection by considering both link and node attributes for social networks, which are the closest to our work. Several existing works on attributed graph clustering fall into this category. The main idea is to design a distance/similarity measure for vertex pairs that combines both structural and attribute information of the nodes. Based on this measure, standard clustering algorithms like K-Medoids and spectral clustering are then applied to cluster the nodes. A weighted adjacency matrix is used as the similarity measure in [23], where the weight of each edge is defined as the number of attribute values shared by the two end nodes. The authors applied graph clustering algorithms on the constructed adjacency matrix to perform clustering. In [24], the authors considered the community structure of the social networks and influence-based closeness centrality measure of the nodes to identify the most influential nodes. The state-of-the-art distance-based approach is the SA-Cluster proposed by Zhou et al. [25] which defined a unified distance measure to combine structural and attribute similarities. Attribute nodes and edges are added to the original graph to connect nodes which share attribute values, and a neighborhood random walk model is used to measure the node closeness on the augmented graph. Afterwards, a clustering algorithm SA-Cluster is proposed based on the K-Medoids method. In addition, modularity-based methods are very popular recently; Duan et al. [26] connected modularity-based methods with correlation analysis to solve the resolution limit problem.

However, all these works in the last category attempted to optimize two contradictory objective functions and intended to identify disjoint communities, and thus the communities detected were not optimal and had no clear semantic meanings. In this work, we propose to leverage both the structure links between users and venues and their attributes to discover the overlapping community structure. Specifically, we formulate the overlapping community detection problem into a multimode multi-attribute edge-centric co-clustering issue, viewing both inter-mode links and intra-mode attributes as unified features for clustering. With this novel representation, users and venues together with their attributes are grouped in a natural way, where the detected communities have explicit semantic meanings that can be interpreted as community profiles.

7.1.3 Problem Statement

Homophily is one of the important reasons that people connect with others [27], which can be observed everywhere in LBSNs. For example, users coming from the same country might have common habits which lead to similar check-in patterns, users who have similar social roles are more likely to check in the same places, and users who check in the same places might have similar preferences and interests. The homophily phenomenon in LBSNs suggests that communities should consist of like-minded users; therefore, a community in LBSNs should be a group of users who are more similar with users within the group than users outside the group. Specifically, the similarity is mainly reflected in people's lifestyles, social roles, preferences, and interests, rather than online inter-user links.

In LBSNs, users tend to frequently check in venues which are important to them or they are interested in. By analyzing the check-in history, we can obtain the user's check-in pattern from temporal, spatial, as well as social aspects, which can be seen as a representation of his/her lifestyle, social role, preferences, or interests. In this work, a community is defined as a cluster of edges (check-ins) with user and venue as two modes, where the common attributes of users and venues characterize the properties of the community.

We use $U = (u_1, u_2, \ldots, u_m)$ to represent the user set, and $V = (v_1, v_2, \ldots, v_n)$ to denote the venue category set; a community $C_i (1 \leq i \leq k)$ is a subset of users and venue categories, where k is the number of communities. On the one hand, the check-in relationship between users and venue categories forms a matrix M, where each entry $M_{ij} \in [0,1)$ corresponds to the number of check-ins that u_i has performed over v_j. Therefore, each user can be represented as a vector of venue categories, and each venue category can be denoted as a vector of users. On the other hand, users and venue categories might have several independent attributes, denoted as $(a_{i1}, a_{i2}, \ldots, a_{ix})$ and $(b_{j1}, b_{j2}, \ldots, b_{jy})$, respectively. Normally, every attribute reveals a certain social aspect of users or venue categories. For instance, a user has a certain number of followers and followings, and a venue category has a common operating time. Therefore, both user mode and venue mode have two types of representations: an inter-mode representation as well as an intra-mode representation.

Based on the above notations, the overlapping community detection in LBSNs can be formulated as a multimode multi-attribute edge-centric co-clustering problem as follows:

Input:

- A check-in matrix $M(|U| \times |V|)$, where $|U|$ and $|V|$ are the numbers of users and venue categories, respectively
- A user attributes matrix $M(|U| \times |A|)$, where $|A|$ is the number of user attributes
- A venue category attributes matrix $M(|V| \times |B|)$, where $|B|$ is the number of venue category attributes
- The number of communities k

Output:

- K overlapping communities which consist of both users and venue categories according to an application-oriented optimization function.

7.1.4   Multimode Multi-Attribute Edge-Centric Co-clustering Framework

The observation that a check-in on LBSNs reflects a certain aspect of the user's preferences or interests enlightens us to cluster edges instead of nodes, as the detected clusters of check-ins will naturally assign users into overlapping communities with connections to venues. Specifically, after obtaining edge clusters, overlapping communities of users can be recovered by replacing each edge with its vertices; that is, a user is involved in a community as long as any of his/her check-ins falls into the community. In such a way, the obtained communities are usually highly overlapped. The key idea of the proposed framework is shown in Fig. 7.2.

As indicated in Fig. 7.2, we first select features based on the characteristics of the collected LBSN dataset and then perform feature normalization and fusion. Second, the overlapping community structure is detected by using the proposed edge-centric co-clustering algorithm, where an edge-cutting mechanism has been introduced to eliminate trivial community memberships. Finally, by combining the detected communities together with user-venue metadata (e.g., user social statues, venue categories), we obtain the community profiles to interpret the social and semantic meanings of communities.

7.1.4.1   Edge-Centric Co-clustering

As stated in the introduction section, we define a community in LBSNs as a group of users who are more similar with users within the group than users outside the group.

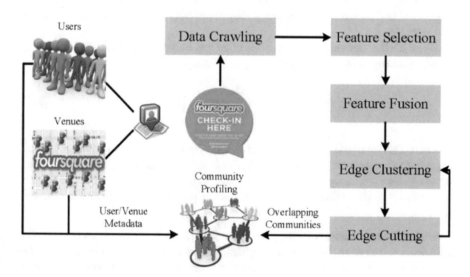

Fig. 7.2 Community discovering framework

Therefore, communities that aggregate similar users and venues together should be detected by maximizing intra-cluster similarity rather than maximizing modularity. This objective function is formulated as

$$\text{OBJ} = \arg \max_{C} \sum_{j=1}^{k} \sum_{e_x \in C_j} \text{sim}(e_x, C_j), \tag{7.1}$$

where k is the number of communities, $C = \{C_1, C_2, \ldots, C_k\}$ is the detected community set, e_c denotes an edge of community C_j, and $\text{sim}(e_c, C_j)$ is the similarity between e_c and C_j. With this formulation, the key is to characterize the similarity between an edge and a community. To this end, we first introduce the definition of edge similarity.

In a user-venue check-in network, each edge is associated with a user vertex and a venue vertex. By taking an edge-centric view, each edge is treated as an instance with its two vertices as features. In other words, the similarity between a pair of edges can be defined as the similarity between the corresponding user pair and venue pair as

$$\text{sim}_{\text{edge}}(e_x, e_y) = \mathcal{F}(\text{sim}_u(u_x, u_y), \text{sim}_v(v_x, v_y)), \tag{7.2}$$

where $\text{sim}_u(u_i, u_j)$ is the similarity between two users, $\text{sim}_v(v_i, v_j)$ is the similarity between two venues, and $\mathcal{F}$ represents the function used to combine these two similarities, by balancing the weights of the user mode and the venue mode. The formalism of $\mathcal{F}$ depends on the characteristics of the expected communities as well as the targeted applications. Considering the similarity trade-off between user mode and venue mode, two widely used formalisms of $\mathcal{F}$ are $(\text{sim}_u + \text{sim}_v)/2$ and $(\text{sim}_u \times \text{sim}_v)^{1/2}$. In this work, we adopt the second notion to ensure that a pair of edges is of high similarity *if and only if* they are of high similarity in both user mode and venue mode.

Since each community contains a set of edges, based on Eq. (7.2), the similarity between an edge e_i and a community C_j is defined as

$$\text{sim}_{e_x, C_j} = \frac{1}{|C_j|} \sum_{e_y \in C_j} \text{sim}_{\text{edge}}(e_x, e_y), \tag{7.3}$$

where $|C_j|$ refers to the number of edges within community C_j and e_c represents an edge from C_j. Obviously, the edge similarity is defined based on two mode similarities (i.e., user similarity and venue similarity). In the following two subsections, we compute the mode similarity by taking into account both inter-mode and intra-mode features.

Inter/Intra-mode Features The inter-mode feature describes the structure similarity between a pair of edges based on the physical links between users and venues, and the intra-mode feature depicts attribute similarity where each attribute corresponds to a certain social aspect of users or venues. As we have mentioned, in many real applications, both inter-mode links and intra-mode attributes are important.

Considering the characteristic of user-venue links (mainly check-ins) in LBSNs, we study two inter-mode features: (1) characterizing a user based on a vector of venue categories, namely *user-venue similarity*, and (2) characterizing a venue category by using a vector of users, which is defined as *venue-user similarity*.

Meanwhile, by analyzing the available user-venue-related metadata in LBSNs, we identify three intra-mode features which are *user social status similarity*, *user geo-span similarity*, and *venue temporal similarity*.

All the abovementioned features will be presented in detail in the empirical study section.

Feature Fusion Due to the characteristic of various similarity features, different calculation methods might be used which lead to different value ranges. Therefore, the absolute values of different features must be normalized. To this end, we simply normalize each similarity measure sim_x into the interval $[0, 1]$ as follows:

$$\text{sim}'_x = \frac{\text{sim}_x - \min\left(\text{sim}_x\right)}{\max\left(\text{sim}_x\right) - \min\left(\text{sim}_x\right)}, \tag{7.4}$$

where sim'_x is the normalized format of measure sim_x. Afterwards, another issue is to fuse different features. Considering that each edge consists of two nodes, we first defined user similarity and venue similarity as

$$\text{sim}_u = \frac{1}{|f_u|} \sum \text{sim}'_{u*}, \tag{7.5}$$

$$\text{sim}_v = \frac{1}{|f_v|} \sum \text{sim}'_{v*}. \tag{7.6}$$

where $|f_u|$ and $|f_v|$ represent the number of selected features for user mode and venue mode, respectively; sim'_{u*} and sim'_{v*} refer to the normalized similarity. Then, based on Eq. (7.2), the edge similarity is calculated as follows:

$$\text{sim}_{\text{edge}} = \sqrt{\text{sim}_u \times \text{sim}_v}. \tag{7.7}$$

Clustering Algorithm Based on the above formulation, the multimode multi-attribute edge clustering problem is converted into an ordinary clustering issue, which can be handled by adjusting k-means as follows:

- While k-means selects the mean (i.e., geometric center) of all the instances (i.e., edges) in a cluster as its centroid, we represent each centroid by using the whole set of instances within the cluster. According to the definition of the similarity between an edge and a cluster, if a set of multimode multi-attribute edges are denoted by a single vector, the obtained similarity will be significantly different.
- The similarity between a given pair of instances is not directly calculated but based on the similarity between the corresponding pair of vertices. As each edge includes two vertices and each vertex consists of multiple attributes which are

usually represented as feature vectors of different dimensions (i.e., length), it is difficult to define a unified distance measure to characterize the similarity between a pair of multimode multi-attribute edges.

- While representing each centroid as a set of instances ensures the precision of the obtained similarity, the computation complexity increases from $O(k \times N)$ to $O(N_2)$. To improve the time efficiency, each centroid C_j is denoted as a structure which consists of four components: a list of current instances within the centroid (E_{C_j}), a list of instances that are assigned to the centroid during last iteration (E_{A,C_j}), a list of instances that are removed from the centroid during last iteration (E_{R,C_j}), and the similarity array between the previous centroid and the whole set of instances ($\text{sim}(E_{P,C_j}, E)$).

Based on the above adjustments, the proposed clustering method is presented in Algorithm 7.1.

Algorithm 7.1: Edge-Centric Co-clustering

Inputs: E, a list of multi-mode edges $\{e_i | 1 \leq i \leq N\}$;
M_u, the user-mode similarity matrix;
M_v, the venue-mode similarity matrix;
k, the number of communities;
Outputs: C, a list of communities $\{C_j | 1 \leq j \leq k\}$;
1: k edges are randomly selected $\{e_j | 1 \leq j \leq k\}$;
2: **for** each e_j **do**
3:      $E_{C_j} \leftarrow \{e_j\}$;
4:      $E_{A,C_j} \leftarrow E_{C_j}$;
5:      $E_{R,C_j} \leftarrow \varnothing$;
6:      $\text{sim}(E_{P,C_j}, E) \leftarrow zeros(|E|)$;
7: **end for**
8: $\{\text{maxsim}_i | 1 \leq i \leq N\} \leftarrow 0$;
9: $\Delta \leftarrow \infty$;
10: **while** $\Delta \geq \varepsilon$ **do**
11:      $\text{Obj}_{pre} \leftarrow \sum \text{maxsim}_i$;
12:      $\{\text{maxsim}_i | 1 \leq i \leq N\} \leftarrow 0$;
13:      **for** each centroid C_j **do**
14:          **for** each $e_i \in E$ **do**
15:              calculate $\text{sim}(E_{A,C_j}, e_i)$;
16:              calculate $\text{sim}(E_{R,C_j}, e_i)$;
17:              $\text{sim}(E_{C_j}, e_i) \leftarrow \text{sim}(E_{P,C_j}, e_i) + \text{sim}(E_{A,C_j}, e_i) - \text{sim}(E_{R,C_j}, e_i)$;
18:              **if** $\text{sim}(E_{C_j}, e_i) > \text{maxsim}_i$ **do**
19 : $\text{maxsim}_i \leftarrow \text{sim}(E_{C_j}, e_i)$;
20:                  assign e_i to C_j;
21:              **end if**
22:          **end for**
23:      **end for**
24:      update the centroids;
25:      $\text{Obj}_{cur} \leftarrow \sum \text{maxsim}_i$;
26:      $\Delta \leftarrow abs(\text{Obj}_{cur} - \text{Obj}_{pre})$;
27: **end while**

At the beginning, k edges are randomly selected (line 1) based on which a set of initial centroids are constructed (lines 2–7). Afterwards, during the iteration, given a centroid C_j we compute the similarity that each edge e_i has obtained (line 15) as well as the similarity it has lost (line 16) during the last reassignment, based on which the current similarity between e_j and C_j is calculated (line 17). Edge e_i is assigned to the centroid that is most similar to itself, and the corresponding similarity is marked as maxsim$_i$ (lines 18–21). Centroid updating is performed based on the reassignment of edges (line 24). At the end of each iteration, the current value of the objective function is calculated (Obj$_{cur}$, line 25) to compare with the previous value Obj$_{pre}$ (line 26). The iteration terminates if and only if the absolute difference between these two values is smaller than the predefined threshold ε (line 27). Experiments based on our dataset show that the algorithm usually converges within 100 iterations.

7.1.4.2 Edge Cutting

Sometimes, the LBSN users perform occasional check-ins, which do not reflect their preferences or interests and may bring negative effect on community detection. To deal with this issue, we put forward a progressive edge-cutting mechanism by introducing a threshold τ to restrict the minimum participation. This threshold determines whether a user should or should not be a member of a given community. We denote the set of edges corresponding to the check-ins of user u_m as E_m. For each community C_j that u_m belongs to, if the ratio between $E_{m,j}$ and E_m ($E_{m,j}$ is the subset of E_m which falls into C_j) is lower than τ, u_m will be excluded from community C_j and the edges within $E_{m,j}$ will be eliminated.

Experimental results show that edge cutting is able to not only limit the overlapping degree (i.e., the average number of communities that each user belongs to) but also improve the community purity. The detailed performance of edge cutting will be presented in the evaluation section.

7.1.4.3 Optimization Measure

Modularity is a widely adopted measure to evaluate community detection methods which aim to cluster nodes based on inter-mode connection. Specifically, given a dataset the optimal community structure can be discovered by maximizing modularity, since the modularity first increases along with k and achieves its extreme at a specific k and then it starts to decrease. However, as we have mentioned, this work aims to group together similar users by maximizing intra-cluster similarity, where the intra-cluster similarity consistently increases as k increases. Therefore, for similarity-based community detection, it is necessary to propose a new measure to evaluate the quality of the detected community structure, which is not a trivial task.

We deal with this issue based on the following two assumptions. On the one hand, a significant community should consist of a certain number of users who are about equally engaged in the community. To this end, our first assumption is that the size of a good community should not be very small and its members should have similar participation degree. On the other hand, the interests of a significant community should mainly focus on only a few social activities. Therefore, our second assumption is that an effective community in LBSNs should consist of a small number of venue categories. Accordingly, we introduce two different community entropies: community participation entropy and community interest entropy, based on which an optimization function is constructed to assess the quality of the detected communities.

Community Participation Entropy Community participation entropy takes into account both the community size (i.e., number of members) and the relative proportion of each member's participation (i.e., check-in times). A community has high participation entropy if it consists of a large number of users with similar participation degree. Conversely, a community has low participation entropy if it is dominated by several users. Specifically, community participation entropy is defined as

$$H_{P, C_j} = -\sum_{u_i \in C_j} \left(\frac{|u_i|}{|C_j|} \cdot \log \frac{|u_i|}{|C_j|} \right), \tag{7.8}$$

where $|Cj|$ is the number of check-ins within Cj, and $|u_i|$ denotes the number of edges within C_j which is from u_i. Obviously, social interactions are most likely to occur within communities of high participation entropy [28].

Community Interest Entropy Community interest entropy measures the evenness of members' check-in preferences within a community. Specifically, a community corresponds to a cluster of user check-ins, and each check-in is associated to a particular venue category. Therefore, a community C_j can be represented as a vector of venue categories $\langle \omega_{C_j,V_1}, \omega_{C_j,V_2}, \ldots, \omega_{C_j,V_N} \rangle$, where ω_{C_j,V_n} denotes the occurrence count of category V_n within C_j. Accordingly, community interest entropy can be formalized as

$$H_{I, C_j} = -\sum_{n=1}^{N} \frac{\omega_{C_j,V_n}}{|C_j|} \cdot \log \frac{\omega_{C_j,V_n}}{|C_j|}, \tag{7.9}$$

where $|C_j|$ is the number of check-ins within C_j which is equal to $\sum \omega_{C_j,V_n}$. A community will have low interest entropy if it only consists of a small number of venue categories with high occurrence counts; it will have high interest entropy if even occurrence counts spread over a big number of venue categories.

Optimization Function According to the above definitions, given a dataset, both the average community participation entropy and the average community interest entropy will decrease as k increases. Meanwhile, an optimal community structure should consist of communities with high participation entropy and low interest entropy. Therefore, the estimation of community structure is formulated as a multi-objective optimization problem as

$$\begin{cases} \arg\max_k \dfrac{1}{k} \displaystyle\sum_{C_j^k \in \{C^k\}} H_{P,C_j^k} \\[2em] \arg\min_k \dfrac{1}{k} \displaystyle\sum_{C_j^k \in \{C^k\}} H_{I,C_j^k} \end{cases} . \tag{7.10}$$

In the above model, $\{C_k\}$ denotes the set of communities detected when the number of communities is set as k and C_j^k is a member of $\{C_k\}$. To solve this model, we first normalize the two community entropies as

$$H'_{P,C_j^k} = \frac{H_{P,C_j^k} - h_P}{H_P - h_P}, \tag{7.11}$$

and

$$H'_{I,C_j^k} = \frac{H_{I,C_j^k} - h_I}{H_I - h_I}, \tag{7.12}$$

where H_P is the expected maximum community participation entropy, h_P denotes the expected minimum community participation entropy, H_I is the expected maximum community interest entropy, and h_I denotes the expected minimum community interest entropy. Afterwards, the above optimization model can be translated into the following format:

$$\arg\max_k F\left(\mu_{P,k}, \mu_{I,k}\right), s.t. \begin{cases} \dfrac{H_{P,C_j^k} - h_P}{H_P - h_P} \ge \mu_{P,k} \\[2em] \dfrac{H_I - H_{I,C_j^k}}{H_I - h_I} \ge \mu_{I,k} \end{cases}, \tag{7.13}$$

where F represents a function on $\mu_{P,k}$ and $\mu_{I,k}$. Obviously, its optimal solution is also an efficient solution for the first model. To this end, we combine the two objectives in this model by introducing an optimization function as follows:

$$Opt(k) = \frac{H_{P,C_j^k} - h_P}{H_P - h_P} \cdot \frac{H_I - H_{I,C_j^k}}{H_I - h_I}. \tag{7.14}$$

On the one hand, for a given k, the optimization function measures the average quality of the detected communities. On the other hand, this function can also be used to compare different community structures that are discovered under different k-values. Therefore, based on the proposed function, we are able to select an optimal k and consequently reveal the optimal community structure. Detailed evaluation of the proposed optimization function is presented in the following section.

7.1.5 Empirical Study Based on Foursquare

Foursquare is a popular LBSN which has more than 20 million registered users till April 2012, and about 3 million check-ins are performed per day [29]. In this section, an empirical study on the proposed overlapping community detection framework is conducted based on the collected Foursquare dataset.

7.1.5.1 Data Collection

Foursquare API [30] provides limited authorized access for retrieving check-in information; therefore we resort to Twitter streaming API [2, 31] to get the publicly shared check-ins in this work. The data collection started from October 24, 2011, and lasted for 8 weeks, which results in a dataset of more than 12 million check-ins performed by 720,000 users over 3 million venues. Meanwhile, we also crawled metadata related to users and venues, including every user's Twitter profile and every venue's Foursquare profile.

Before community detection, we preprocess the collected dataset as follows. First of all, we excluded check-ins that are performed over invalid venues. In this work, invalid venues refer to those that cannot be resolved by Foursquare API, and thus the detailed information of these venues is not available. Consequently, about 7.52% of the check-ins are removed from the dataset. Secondly, we only keep users who have performed at least one check-in per week on the average (referred as active users), which means inactive users together with their check-ins are excluded. Finally, users who used agent software conducting remote and large-scale automatic check-ins (with a check-in speed faster than 1200 km/h, which is the common airplane speed) are defined as *sudden move* users [12], and check-ins from these users are eliminated. We observed a total number of 9276 *sudden move* users, which occupy about 3.36% of the active users.

After the above data cleansing, the remained dataset includes 266,838 users and 9,803,764 check-ins which were performed over 2,477,122 venues.

7.1.5.2 Feature Description

As abovementioned, in this work we mainly focus on two inter-mode features and three intra-mode features. In this section, we present the detailed description of these features based on the characteristic of the collected Foursquare dataset.

1. *Inter-mode Features*

   User-Venue Similarity: Foursquare classifies venues into 400 categories under 9 parent categories. We identify 274 venue categories by merging those similar ones, and consequently based on a user's check-in venues, each user can be represented as a vector of 274 dimensions. We build a $266,838 \times 274$ matrix to represent all the active users within the collected dataset. Afterwards, this matrix is refined based on principal component analysis, which is able to convert a set of observation of correlated variable into a set of value of linearly uncorrelated variable under a latent space. By applying principal component analysis on the raw matrix, we obtain a $266,838 \times 100$ matrix which covers 95.62% of the total variance. After the conversion, each user is represented as a vector of 100 dimensions in the latent space. Based on the above matrix, the user-venue similarity for a pair of users u_m and u_n is calculated based on cosine similarity.

Venue-User Similarity: As we have mentioned, each venue category of Foursquare can be denoted as a vector by treating users as features as well. Following the same approach as the above section, we obtain a 274×100 matrix by performing principal component analysis on the original $274 \times 266,838$ matrix, which covers 95.34% of the total variance. As a result, each venue category corresponds to a vector of 100 dimensions in the latent space. Similarly, the venue-user similarity is also defined using cosine similarity.

2. *Intra-mode Features*

User Social Status Similarity: There are two lists in each user's Twitter profile, a follower list and a following list. We define a user's social status as the ratio of his/her number of followers to his/her number of followings. Specifically, the social status of a user u_m is formalized as

$$S_s = \frac{n_{\text{followers}}(u_m)}{n_{\text{followings}}(u_m)}. \tag{7.15}$$

According to the above definition, users with high social status are those who have many followers and fewer followings. To some extent, these users act as hubs of the social network.

We introduce the first intra-user similarity feature, namely user social status similarity based on the user social status metric. Given a pair of users u_m and u_n, this feature is defined as

$$\text{sim}_{us}(u_m, u_n) = \frac{\min(s_{s,m}, s_{s,n})}{\max(s_{s,m}, s_{s,n})}, \tag{7.16}$$

where $S_{s,m}$ and $S_{s,n}$ represent the social status of u_m and u_n, respectively. Apparently, the value of user social-status similarity falls into the interval [0, 1].

User Geo-Span Similarity: The user geo-span (a.k.a. radius of gyration) is another metric that can be used to distinguish the lifestyle of different users, which is defined as the standard deviation of distances between a user's check-ins and his/her home location. In LBSNs, a user's home location is defined as the centroid position of his/her most popular check-in region [9]. The user geo-span metric is able to indicate not only how frequently but also how far a user moves. Generally, a user with low radius of gyration mainly travels locally (with few long-distance check-ins), while a user with high radius of gyration has many long-distance check-ins. The formal definition for radius of gyration is as follows:

$$r_{\mathrm{g}} = \sqrt{\frac{1}{n} \sum_{i=1}^{n} (r_i - r_{\mathrm{h}})^2}, \tag{7.17}$$

where n is the number of check-ins made by a user, and $r_i - r_{\mathrm{h}}$ is the distance between a particular check-in location r_i and the user's home location r_{h}.

By using the radius of gyration metric, we introduce the second intra-user similarity feature named user geo-span similarity. Specifically, for a pair of users u_m and u_n, the calculation of this feature is same as the user social status similarity.

Venue Temporal Similarity: Generally, people visit and check in different kind of venues at different times, such that different venue categories can be distinguished according to their temporal check-in patterns [32]. In this work, we divide a week into 168 (7×24) time slots and each time slot corresponds to 1 h in a certain day of the week, reflecting the temporal characteristic of each user check-in. Build a weekly temporal check-in band for each venue category at the hour granularity, which means that each temporal band corresponds to a vector of 168 dimensions (7×24). For example, Fig. 7.3 plots the check-in patterns of two different venue categories: the bar and the museum. We can see that, according to the temporal check-in patterns, museums are most popular during the daytime of weekends while bars are extremely busy on Friday and Saturday evening.

Since we have identified 274 venue categories, a 274×168 matrix is constructed and then principal component analysis is performed on this matrix, producing a new matrix of 274×20 which covers 99.92% of the total variance. Consequently, the venue temporal similarity between a pair of venues can also be defined based on cosine similarity.

Based on the normalization and fusion mechanisms presented in Sect. 7.1.4.1, different forms of edge similarities can be obtained by using different feature combinations, which are able to reveal the community structures in LBSNs from different perspectives.

7.1.6 Performance Evaluation

To evaluate the performance of the proposed framework, we chose three big cities (i.e., Paris, New York, and Tokyo) as the target societies, and collected Foursquare check-ins accordingly. Meanwhile, we designed a set of baseline methods by using different features (as summarized in Table 7.1). For more details about data collection and experimental setup, please refer to the article [33].

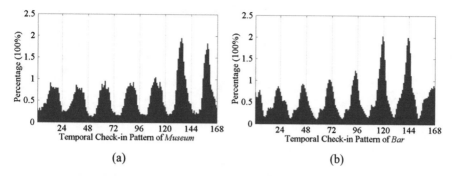

Fig. 7.3 Temporal check-in patterns of two different venues: (**a**) museum; (**b**) bar

Table 7.1   Different feature sets evaluated in the experiments

Feature set	Used features
I	UV (user-venue similarity) and VU (venue-user similarity), which are the same as edge clustering and used as the baseline in this work
II	UV, VU, and VT (venue temporal similarity)
III	UV, US (user social status similarity), VU, and VT
IV	UV, UG (user geo-span similarity), VU, and VT
V	UV, UG, US, VU, and VT

7.1.6.1   Co-clustering Results

The first experiment was designed to evaluate the quality of the detected communities. As we do not know the ground truth community structure [34] of the used Foursquare dataset, we adopted an indirect evaluation approach. Intuitively, users visiting similar venues tend to share similar interests, which are reflected through the topics they discuss (i.e., tips). Therefore, we chose to estimate the proposed community detection method by assessing whether the tips that are posted by the same community are also of high similarity. In other words, an effective community detection method should achieve high *community tip similarity*. We performed community detection using the proposed framework in the Paris dataset, and repeated experiments ten times for each of the five feature sets listed in Table 7.1. The average community tip similarity of different feature sets is shown in Fig. 7.4.

According to Fig. 7.4, we have the following observation. First, *Feature Set IV* is the most competitive feature set while *Feature Set V* is the next most competitive one, where the *user geo-span* feature has been leveraged. This indicates that users who have similar geo-spans are most likely to discuss similar topics. Meanwhile, while most of the introduced intra-mode features can increase the community tip similarity, the *user social status* feature will lower the performance; that is, the performance of *Feature Set III* is worse than the baseline method (i.e., *Feature Set I*). The reason might be that there is no correlation between users' social statuses and

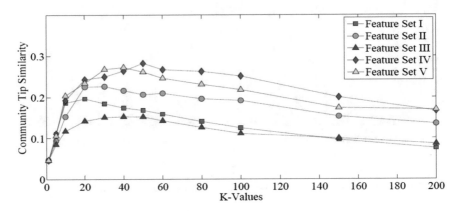

Fig. 7.4 Community tip similarity of different feature sets

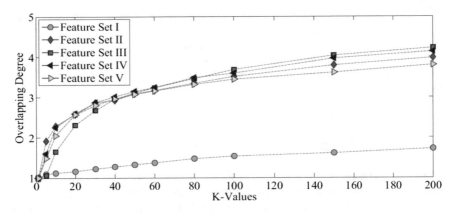

Fig. 7.5 Overlapping degree of different feature sets

their tip topics. Similar results can be obtained based on the Tokyo and New York dataset.

The second experiment was designed to evaluate the proposed framework's ability to uncover overlapping communities; that is, how would different feature sets impact the overlapping degree of the detected communities. The results are shown in Fig. 7.5.

According to Fig. 7.5, community detection methods which have leveraged not only inter-mode but also intra-mode features are able to achieve much higher overlapping degrees. The result implies that intra-mode features can capture certain aspects of human activity patterns that cannot be characterized by inter-mode features, even though a high overlapping degree does not necessarily represent the quality of a community detection method.

We also have designed experiments to evaluate performance of the proposed optimization function and the edge-cutting mechanism. For more details about the experimental results, please refer to the article [33].

7.2   Understanding Social Relationship Evolution by Using Real-World Sensing Data[2]

7.2.1   Introduction

Learning patterns of human behavior from sensory data is an emerging domain aimed at understanding the habits and routines of individuals [35–37]. Besides individual behavior, the relationship between humans is a topic explored widely in social, cognitive, and computer science [38–41]. Determining social relationships is of potential for achieving important societal goals, ranging from epidemic prevention to urban planning and public safety [42].

A social relationship is subject to constant evolution due to frequent changes in individual activities and group interactions [43]. Investigating how a social group evolves, on the one hand, is important to understand community dynamics and predict its future structure. On the other hand, it is challenging due to the difficulty in obtaining dynamic and continuous human interaction information reflecting the evolution process in the real world. While a number of studies of social network analysis have been conducted in online social media [44–46], little has been known for physically connected data. Thanks to mobile and pervasive computing technologies, we are able to capture such interaction data by using ubiquitous sensors such as mobile phones [42, 47].

In contrast to general social networks, we focus on a tight-knit social relationship here, i.e., friendship. One important step in detecting friendship evolution is to predict friendship, especially the nonreciprocal friends. Eagle et al. [42] propose to infer friendship based on proximity (via Bluetooth scan) captured by mobile phone. It is based on the common sense and experience that friends usually spend time together in the same physical sites. By using the extra-role factor, i.e., off-campus proximity, they can predict most reciprocal friends and non-friends. However they analyze friendship with undirected dyads, i.e., *friend(A—B)*, where *A* and *B* constitute a dyad. Hence their approach does not work in determining nonreciprocal friends. The proximity of nonreciprocal friendships ranges from quite small to very large.

7.2.2   Related Work

There has been a number of works done over the past decade to analyze human communication patterns and relationships by using the data directly extracted from the Web [46, 48, 49], email [50, 51], and instant messaging [52]. Tang et al. [46]

---

[2]Part of this section is based on a previous work: Z. Yu, X. Zhou, D. Zhang, G. Schiele, C. Becker. Understanding social relationship evolution by using real-world sensing data. World Wide Web (2013) 16: 749. https://doi.org/10.1007/s11280-012-0189-x.

proposed to detect community based on social media data acquired from the Web, such as BlogCatalog and Flickr. Lin et al. [49] used the Blog and DBLP data to discover communities and analyze community evolution. Chen and Saad [48] adopted the citation network and trust network on the Web to extract community. Kossinets and Watts [50] succeeded to infer interactions between individuals and social network from time-stamped email headers. Malmgren et al. [51] used email records to understand individual communication patterns. Leskovec and Horvitz [52] examined characteristics and patterns that emerge from the collective dynamics of a large number of people by analyzing the data captured from the Microsoft Messenger instant messaging system. Hristova et al. [53] identified human relationships off-line and online by considering the number of communication channels utilized between students.

For its inherent advantages in ubiquity and mobility, mobile phones have been widely adopted to capture data for the analysis of human mobility, communication patterns, and social networks. Barabasi et al. [54, 55] studied individual human mobility patterns by using the trajectories of 100,000 mobile phone users in a 6-month period. They found that human trajectories show a high degree of temporal and spatial regularity that leads to high predictability in user mobility. Eagle [56] used mobile phones as a cultural lens, that is, using mobile phone data captured in the USA, Kenya, Finland, Rwanda, and the UK, respectively, to understand human life patterns with different cultural backgrounds in terms of pace of life, reactions to outlier events, and social financial supports. Wesolowski and Eagle [57] proposed to use data generated from mobile phones to understand the social status of one of the largest poor communities, Kibera in Kenya, specifically focusing on migration patterns, work trends, and tribal affiliations. Onnela et al. [47] analyzed the structure and tie strengths of social and communication networks by using the call records of millions of mobile phones. The Reality Mining project [54] investigated friendship network structure and user satisfaction to their working environments based on mobile phone data. Dong et al. [58] proposed to discover individual friendship pattern from cellular phone call logs. Wang et al. [59] learned region representation using the large-scale taxi flow data and measured the relationship strengths between regions. Zhang et al. [60] employed the residual neural network framework to predict the direction of crowd flows by utilizing taxicab and bike trajectory data. Fan et al. [61] predicted human movements by analyzing the crowd behavior-utilized mobile phone GPS log dataset.

In general, the shared goal of the abovementioned works is to analyze human communication patterns, mobility patterns, communities, and social networks by using directly observable data. Our work differs in several aspects. First, we define friendship as a directed relation which is different from the current undirected researches. Second, our work recognizes human friendship from a supervised learning perspective which has been proved to be more effective in determining nonreciprocal friends than simply judging from a factor score [54]. Third, besides inferring friendship, we also explore the evolution of the friendship network over time by using the real-world sensing data.

Numerous studies have investigated social relationship and social network evolution. They can be classified into three categories based on their data source for analysis: user survey data, agent simulation, and web data. Leenders [62] used survey data from classrooms to examine the effect of gender similarity and reciprocity on friendship and best friendship choices. The results showed that gender similarity is profoundly predominant in the "best friend" networks and "best friend" choices are stable. Han et al. [63] formalized a model of predicting potential invitees of groups considering social groups. Khanafiah et al. [64] applied agent simulation to analyze the interpersonal relationship based on Heider's balance theory, which shows how a social group evolves to a possible balance state. Barabasi et al. [65] investigated the evolution of social networks by using the co-authorship network built based on a web database including all relevant journals in mathematics and neuroscience for an 8-year period. Kumar et al. [62] and Leskovec et al. [66] studied the evolution of large online social networks based on the dataset extracted from Flickr, Yahoo! 360, Yahoo! Answers, Delicious, and LinkedIn. All the above researches do not use real-world sensing data. Liu et al. [67] examined the evolution of the research field by applying graph theory and co-authorship network. Palla et al. [43] investigated the stability, group lifetime, and member abandonment in social group evolution by using both co-authorship network and mobile phone call records. Kossinets and Watts [50] analyzed social network evolution by using email contact data and found that network evolution is dominated by the network topology and the organizational structure in which the network is embedded. However both of the two studies are not focused on friendship evolution.

7.2.3   Friendship Prediction

Before talking about social relationship evolution, we need to investigate how a social relationship can be quantified and predicted. We propose an approach for predicting friendship by using mobile phone data captured in real life. Hence we have the first hypothesis:

Hypothesis 1 *Social relationships (e.g., friends and non-friends) can be predicted by using real-world sensing data.* Several existing studies attempt to quantify human relationships by using directly observable data, such as email [50, 68], instant messaging [52], and phone records [47].

Assuming that friendship can be predicted, we would like to examine the difficulty in predicting different types of friendships, i.e., nonreciprocal friends and reciprocal friends. Hypothesis 2 refers to this.

Hypothesis 2 *In terms of friendship prediction, nonreciprocal friends are more difficult than reciprocal friends.* Other research has indicated that nonreciprocal friendships are indicative of moderately valued friendship ties [42].

7.2.3.1 Features

To infer whether a user regards the other user as his/her friend, we use the following five features: proximity, outgoing calls, outgoing text messages, incoming calls, and incoming text messages. These features are extracted from the MIT mobile phone data.

Proximity It is intuitive that friends usually spend time together in the same physical sites. Proximity—especially off-campus on weekends and evenings—has been verified to be a good indicator for friendship [42]. It is represented and quantified as the extra-role factor score, i.e., the times of being together off-campus on weekends and evenings in a period. We also choose this feature in our system. Actually the factor encapsulates three aspects: location (on-campus or off-campus), time (weekday or weekend, daytime or evening), and proximity. The location can be inferred from the cellular tower transition data. The proximity is detected by Bluetooth scans. The time information is directly extracted from the time stamp of each Bluetooth scan.

Outgoing Calls and Outgoing Text Messages Initiating communication frequently with a person often implies that the initiator regards the receiver as his/her friend. For example, when a person faces a problem, he/she is likely to contact his/her friends for a favor. The feature of outgoing calls is measured as the sum of duration of voice calls to a particular user in the observation period. We define the feature of outgoing text messages as the total number of sent messages by a user to a particular user during the observational period. The two features are obtained through the statistics of the call logs.

Incoming Calls and Incoming Text Messages Receiving calls or messages from a person is also an important determinant of friendship. It is intuitive that a person should often receive communications from the ones whom he/she regards as friends. Similarly, the feature of incoming calls is calculated as the summed-up total duration of all voice calls from a particular user in a period. The feature of incoming text messages is simply the total number of the short messages the user receives from another user in the observation period. These two features are also extracted from the communication logs.

To examine the effect of different communication directions and types in friendship prediction, we have another two hypotheses as follows:

Hypothesis 3 The outgoing communications play a more important role than the incoming ones in predicting friendship. One would expect that the outgoing (active) communications are more important than the incoming (passive) communications in maintaining his/her friendship network.

Hypothesis 4 *Voice calls are more useful than text messages in friendship determination.* Usually voice calls occupy more proportion in human beings' daily communications that should be more important in friendship determination.

7.2.3.2 Inference Model

This model adopts the SVM classifier for friendship recognition. Given the features of an instance, the SVM classifier sorts it into one of the two classes of relationships: friend and non-friend. SVM has been proven to be powerful in classification problems, and often achieves higher accuracy than other pattern recognition techniques [69]. It is well known for its strong ability to separate hyperplanes of two or more dimensions.

Specifically, this system uses the LIBSVM library [70] for classifier implementation. The default kernel type is the radial basis function (RBF): $K(x_i, x_j) = \exp(-\gamma \|x_i - x_j\|^2)$, $\gamma > 0$. Two parameters are important while using LIBSVM: C and Y. Good (C,Y) enables the classifier to accurately predict unknown data (i.e., testing data). In this system, we adopt a "grid search" to find the best parameter C and Y using cross-validation. Basically pairs of (C,Y) are tried and the one with the best cross-validation accuracy is picked.

7.2.4 Social Relationship Evolution

According to Heider's social balance theory [71], human beings generally prefer a balanced and harmonious state in their social environments. If there are any imbalanced structures in their social network, human beings will be of tension and pressure, which will therefore facilitate the network to evolve to a balanced state.

There are two basic principles for a balanced friendship network. One is reciprocality; that is, when a user A regards the other user B as his/her friends, he/she also expects user B to regard him/her as a friend. Hallinan [72] also indicated that nonsymmetric friendships are less stable than symmetric ones. They exist but are only a temporary condition, and in the near future will evolve to symmetry. Figure 7.6 depicts this case, i.e., the evolution process of nonreciprocal friends to reciprocal friends. From the perspective of reciprocality of social balance theory, we have the following hypothesis:

Hypothesis 5 *According to the reciprocality in the social balance theory, the nonreciprocal friend relations are likely to evolve to reciprocal friends.*

The other basic feature of a balanced friendship network is transitivity; that is, when user A and user B are friends, and user B has another friend user C, then user A and user C could be friends with high probability. In other words, user B acts as the medium of friendship between user A and user C. Actually we human beings make most friends in such a way in our daily lives. Transitivity should plan an important role in social relationship evolution as demonstrated in Fig. 7.7. To understand social relationship evolution from the perspective of transitivity of social balance theory, we have the following hypothesis:

Fig. 7.6 Evolving from
nonreciprocal friends to
reciprocal friends

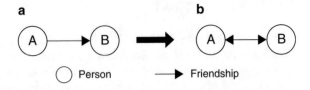

Fig. 7.7 Social relationship
evolution according to the
transitivity

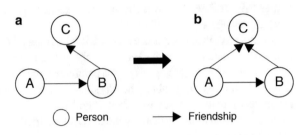

Table 7.2 Prediction accuracy

Relationship	Accuracy by number	Class average accuracy
Friend	0.818	–
Non-friend	0.977	–
Total	0.970	0.898

Hypothesis 6 *The transitivity in the social balance theory should be reflected in the*
social relationship evolution process.

7.2.5   Performance Evaluation

In this section, we present our experimental results based on the MIT Reality Mining
dataset. For more details about the experimental setup, please refer to the article [73].

7.2.5.1   Results of Friendship Prediction

The first experiment was designed to evaluate the performance of friendship prediction. We used all the five features (proximity, outgoing calls, outgoing text messages, incoming calls, and incoming text messages), and the results were
summarized in Table 7.2. Our system achieved a recognition rate of 97.0% by
number and a class average accuracy of 89.8%. Thus, Hypothesis 1 is demonstrated
to be true.

We further examine the difference in recognizing the two types of friendships,
i.e., reciprocal friends and nonreciprocal friends. As shown in Table 7.3, we can find
that the performance of the reciprocal friends is better than that of the nonreciprocal
friends. The reason might be that the features of friendship are more salient with the

Table 7.3 Prediction accuracy for different types of friendships

Friendship type	Accuracy by number
Reciprocal friend	0.857
Nonreciprocal friend	0.750

reciprocal friends than the nonreciprocal friends. This finding also verified that inferring nonreciprocal friendship is more difficult than reciprocal ones due to the fact that nonreciprocal friendships are indicative of moderately valued friendship ties [42]. Thereby, Hypothesis 2 is confirmed.

To examine the difference of the communication directions in determining friendships, we configured two feature subsets (outgoing calls and messages vs. incoming calls and messages) and fed these into the SVM model. In terms of the communication direction, one would expect that the outgoing (active) communications play a more important role in determining the friendship than the incoming (passive) ones. Results in Table 7.4 verified this expectation, and Hypothesis 3 is confirmed.

Moreover, to investigate the difference of the communication types in determining friendships, we combined another two feature subsets (voice calls vs. short messages) and fed them into the SVM model. According to the results shown in Table 7.5, we can find that voice calls are more effective than text messages in friendship determination, which confirmed Hypothesis 4.

7.2.5.2   Results of Friendship Evolution

As the self-report survey of friendship of the MIT dataset was conducted in January 2005 and the data collection was ended in May 2005, we decided to use the data in this period to investigate the evolution of friendship.

The probability of evolution from friend to non-friend is very small since it is almost impossible you forget someone that you have known previously in your group, while the probability of evolution from non-friend to friend is very high as people in a group are encouraged to meet and know each other [64]. Thereby, we mainly focused on the evolution from non-friend to friend, and investigated our two hypotheses about social relationship evolution.

We adopted the training model learned in the previous section to predict the 870 relationships in May 2005, and set the observational period of statistics of the features as January 2005 to May 2005. The predicted results are depicted in Fig. 7.8. Specifically, the self-reported friendship network in January 2005 was demonstrated in Fig. 7.8a, where nodes stand for subjects and arrows represent that one user (start of the arrow) regards the other user (end of the arrow) as his/her friend. The numbers associated with the nodes are user IDs in the MIT Reality Mining dataset [42]. Figure 7.8b depicts the evolved friendship network (in May 2005) by adding the predicted friendships on the basis of self-reported ones. We can find that there were a total of ten friend relations evolved from non-friends.

First, among the 10 evolved friend relations, 7 of them ($6 \rightarrow 2$, $6 \rightarrow 20$, $11 \rightarrow 2$, $23 \rightarrow 8$, $25 \rightarrow 56$, $50 \rightarrow 20$, and $79 \rightarrow 10$) comply with the principle of reciprocality

Table 7.4 Effect of different communication directions

Feature sets	Accuracy by number (friend)	Accuracy by number (non-friend)	Accuracy by number (overall)	Class average accuracy
Outgoing communications (outgoing calls and text messages)	0.667	0.980	0.966	0.824
Incoming communications (incoming calls and text messages)	0.545	0.982	0.961	0.764

Table 7.5 Effect of different communication types

Feature sets	Accuracy by number (friend)	Accuracy by number (non-friend)	Accuracy by number (overall)	Class average accuracy
Voice calls (outgoing calls and incoming calls)	0.576	0.985	0.966	0.781
Short messages (outgoing messages and incoming messages)	0.212	0.995	0.958	0.604

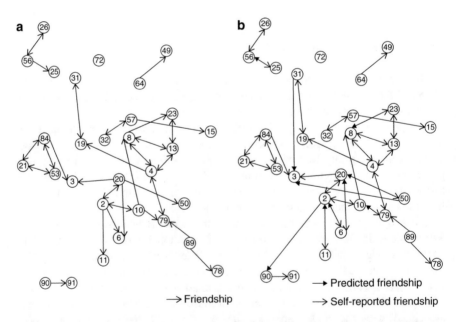

Fig. 7.8 Friendship network and its evolution. (**a**) Self-reported friendship network in January 2005; (**b**) evolved friendship network in May 2005 (plus predicted friends from non-friends)

in the social balance theory. In other words, most of the friendship evolutions are from nonreciprocal friends to reciprocal friends, which confirmed Hypothesis 5.

Second, we can observe that the non-friend relation of user 50 to user 3 changes to a friend relation. It can be interpreted as "user 50 likes user 20 and user 20 likes user 3, then user 50 likes user 3." Obviously it complies with the principle of transitivity in the social balance theory. Furthermore, the evolved friend relations, $6 \rightarrow 2$, $6 \rightarrow 20$, and $23 \rightarrow 8$, also follow the transitivity principle. As a result, we can conclude that the transitivity of the social balance theory is reflected in the social relationship evolution process, and Hypothesis 6 is confirmed.

Moreover, among the 10 evolved friend relations, there are two cases ($2 \rightarrow 90$ and $31 \rightarrow 3$) that do not follow the principles of reciprocality or transitivity. This is understandable as there are really friendships that are not built with someone that we know before or through other friends in our daily lives. We make such friendships spontaneously and serendipitously. These complementary phenomena to social balance theory often lead to the freshness and diversity of different people's friendship networks.

7.3 Interlinking Off-Line and Online Communities[3]

7.3.1 Introduction

People now connect and interact in heterogeneous social communities within cyber-physical spaces. Two forms of social communities have gained particular popularity in recent years. The first type is online communities in the virtual world, where people are connected by sharing content, opinions, and experiences in Web-based social network services (e.g., Facebook, Twitter). The second type is opportunistic communities in the physical world [74–78], which exploits opportunistic contacting and ad hoc connection between pairs of devices (e.g., mobile phones, vehicles) to disseminate information (e.g., local traffic information, public profile). It mimics the way people seek information via social networking through direct, face-to-face contacts.

The two forms of communities have distinct features. First, *they have distinct formation manners and exist in separate spaces*. Online communities enable people to foster interaction with their friends irrespective of physical distance in the virtual space, while opportunistic communities pertain to a traditional way of interaction to people within physical proximity. Second, *they differ from each other on infrastructure support and work environments*. Online communities rely on services in the network infrastructure, and can be accessed by people in the environments with Internet connection (e.g., at home, in the office, hot spots with wireless access

---

[3]Part of this section is based on a previous work: B. Guo, Z. Yu, X. Zhou and D. Zhang, "Hybrid SN: Interlinking Opportunistic and Online Communities to Augment Information Dissemination," The 9th International Conference on Ubiquitous Intelligence and Computing, Fukuoka, 2012, pp. 188–195. DOI: https://doi.org/10.1109/UIC-ATC.2012.29.

points). Opportunistic communities are developed based on mobile ad hoc networks (infrastructure-less), where co-located users can build connections leveraging short-range radio techniques (e.g., Bluetooth). They have advantages in connecting and providing collaboration services to traveling users, who have difficulty in connecting the network infrastructure [74, 75, 78]. Although there are many differences, the two forms of communities also share some commons. For example, they all facilitate information dissemination and social collaboration among users. Further, people are involved in both forms of communities, and they often switch their roles among them (or co-present in one or more communities) in their daily life.

In the past few years, significant research efforts have been made in facilitating information sharing in both online and opportunistic communities. However, they follow separate research lines, and the interlinking of the two forms of communities has little been explored (see Sect. 7.2). In this work, the *hybrid social networking* (HSN) infrastructure is proposed. It is inspired by the multi-community involvement and cross-community traversing nature of modern people. For example, at one moment, Bob is staying at a place with Internet connection and he can communicate with his online friends (in the *online community*); later, he may travel by train with merely ad hoc connection with nearby passengers (forming an *opportunistic community*). Here we use HSN to indicate the smooth switch and collaboration between online and opportunistic communities. By aggregating the complementary features of both networks, HSN can enhance the efficiency and population coverage of information dissemination in comparison with single-network environments. We illustrate the concept of HSN through the following "*opportunistic trading*" use case.

Bob has a "Harry Potter" book, and he wants to sell it by using OBuy. OBuy is a HSN service that supports sell/buy request dissemination & matchmaking across heterogeneous communities. Bob's request is firstly published and shared in the **online community** *(a social network site). If no online-match exist, it is disseminated in opportunistic networks by exploiting the opportunistic contact of OBuy users. To augment the dissemination process in opportunistic networks, several friends of Bob from the* **online community** *are chosen as "brokers" (to carry and forward Bob's request). When Bob and the selected brokers move in the physical world, the request is replicated to physically adjacent people (forming* **opportunistic communities***) and matched. Once a match is detected, the potential "buyer" is informed and the transaction can then be conducted.*

The above use case demonstrates the floating of information over heterogeneous communities.

7.3.2 Related Work

This study brings together research on social connection and information sharing/dissemination in communities, spanning two major directions: online communities and opportunistic communities.

Online communities exist in many forms in the virtual world, such as those based on friendship (e.g., Facebook), profile similarity or common interests (e.g., LinkedIn, Twitter, YouTube), and location dependence (e.g., Foursquare). Individually each form indicates a certain type of connection or similarity between people [79]. In other words, people in online communities are linked based on their existing social relationships or feature similarity. When people connect in online communities, they can share information with their social links. For example, LinkedIn allows users to share their profile and work experience with their directed links. The mobile social network application CenseMe [80] exploits smartphones to automatically infer people's presence (e.g., walking on the street) and then shares this presence with their friends. However, since the linkage relations in online communities are mostly static, the "audience" covered for the information shared is usually limited. Chiu et al. [81] integrated social support and social identity theory to examine factors affecting citizenship behaviors in online support communities. Kim and Hastak [82] explored patterns created by the aggregated interactions of online users on Facebook during disaster responses.

Opportunistic communities are formed spontaneously (in an ad hoc manner) in the physical world, following the way humans come into contact. Unlike online communities, people in opportunistic communities are connected based on chiefly physical proximity. The members of an opportunistic community can share content with each other, but consumers of a content may not be involved at the same time in the same community (opportunistic communities are not formed based on common interests). Benefiting from its human-centric nature, information from an opportunistic community can be opportunistically disseminated (to potential interested users) when its carriers move (from one opportunistic community to another) [74]. However, this usually comes at the price of additional delay in information delivery. Therefore, many studies have been conducted to achieve efficient information dissemination in opportunistic networking environments. A trivial solution is epidemic-like flooding, where all users contacted become the so-called brokers to carry and forward the shared information [83, 84], but this would clearly saturate both network resources (in terms of available bandwidth) and device resources (e.g., in terms of energy, storage, and so on). A better solution is to replicate the information to only "selected" nodes that have more chances to contact and influence others. To this end, researchers start to explore *mobility patterns* [77, 85, 86] as key pieces of context information to predict nodes' activeness and estimate their "social popularity" to serve as brokers. Alim et al. [87] explored the structural vulnerability of social based forwarding and routing methods in opportunistic networks. Tao et al. [88] exploited community structure for opportunistic forwarding decisions in mobile social networks and found that people will have closer relationships and more opportunities to contact with each other if they are in the same community. Zhu et al. [89] studied opportunistic MSN architecture and investigated the social metrics from encounter, social features, and social properties, respectively.

As reviewed above, research on online and opportunistic communities follows two separate research lines. The interlinking of the two forms of communities has yet little been explored. There have been studies about social network analysis across heterogeneous networks. For example, Tang et al. [90] developed a framework for classifying the type of social relationships by learning across different networks (e.g., email network, mobile communication network). Researchers from CMU study the relationship between the users' mobility patterns and structural properties of the online social network, to identify the implicit social link between physical interaction and online connection [91]. However, to the best of our knowledge, the combination of the complementary features of distinct social networks to augment information dissemination remains unstudied.

Rather than viewing online communities and physical communities as competing entities, we think of them as complementary. In view of the cross-space presence and multi-community involvement nature of modern people, HSN is designed in a way that links virtual and physical social networks to enhance both. By combining the merits of both networks, HSN can (1) enlarge the population covered to a shared message in comparison with online social networks and (2) decrease the latency of information dissemination compared with purely opportunistic networks. Particularly, we demonstrate how HSN augments information dissemination through the opportunistic trading use case.

7.3.3   An Overview of HSN

This section provides necessary background information required to present the main contribution of this study by presenting HSN-enhanced information dissemination and giving the architecture of HSN.

7.3.3.1   HSN-Enhanced Information Dissemination

Information dissemination in HSN can be formulated as follows. Users are by default involved in a global, online community C_{on}. When people moves, opportunistic contacts (co-located users) can form a series of opportunistic communities $\{C_{opp}(1 \sim n)\}$. Any user involved in a community is its member, and a user (e.g., u_i) often switches his/her membership while he/she moves (e.g., $u_i \in C_{opp}(i) \rightarrow u_i \in C_{opp}(j)$).

The information shared has two major types: content and request. Content dissemination refers to replication of the content from one node to another. Request dissemination has an additional requirement on request matchmaking (e.g., to match the buy/sell request in the opportunistic trading use case). In the following, we focus on presenting request dissemination. Requests from community members are shared within a community, which forms the request pool of this community, formulated as $C(i)$.req. For a given request ri, we do not assume any particular destination of it; rather, the request is disseminated over and across the communities with a matched request of shared interests that are the potential destinations of the message. In HSN, we say that request $r1$ matches $r2$ when the degree of similarity is above a predefined threshold δ; that is, $sim(r1, r2) > \delta$. Note that each request must have a time to live

(TTL) [77, 85], before which the publisher of this request expects to receive at least one matched user. The request expires if no matched result is obtained during this period. For example, the sell request in the opportunistic trading use case can be set to 1 week.

In opportunistic network studies, to facilitate information dissemination, brokers are often used to carry and forward information, such as epidemic brokers [83, 84] or popular brokers (to select the nodes who are more likely to meet the largest number of people) [77, 92]. One basic assumption is used here, which considers that all users are willing to act as brokers (the so-called selflessness brokers). However, traditional broker-based protocols have several limitations. (1) The selflessness assumption does not always hold. According to social theories, most people are selfish [93, 94]. Since brokers have to contribute computational resources during the data carrying and forwarding process, a node may not be willing to forward packets for others. Therefore, previous algorithms may not work well since some packets are forwarded to nodes unwilling to relay, and will be dropped (the "data drop" issue). (2) In existing popularity-based broker selection protocols [77, 92], brokers are chosen based on direct contact of popular nodes (whose popularity is higher than a predefined threshold). However, due to the dynamic feature of human mobility, popular nodes may not be chosen if the publisher does not meet them. Besides, it also suffers from high delay on broker selection. (3) Existing opportunistic information dissemination protocols suffer from problems on task notification and termination. Due to dynamic network topology, when a matched user is found, how to notify the result to the publisher becomes a big challenge. Also, this result should also be informed to other brokers to terminate their dissemination task. The encounter history-based routing mechanism has been used to address this issue [85]. However, it suffers from the delivery failure problem due to the dynamic network topology. Moreover, the performance of routing-based approaches largely decreases when more brokers are recruited (and should be notified). The interlinking of online and opportunistic communities in HSN brings new opportunities to address the above three issues.

Social Willingness-Based Broker Filtering To address the selfishness issue, existing studies have focused on using reputation-based approaches [95, 96]. We, instead, want to capture user selfishness in a more realistic manner, mainly from the social perspective. It has been observed that a selfish user is usually willing to help others with whom he/she has social ties (e.g., friends, coworkers, roommates), because he/she gets help from them in the past or will probably get help from them in the future [97]. We call it the *social willingness* phenomena. The broker selection protocol of HSN is founded on this phenomenon, where a request from a publisher can be copied to a selected number of his/her acquaintances to release the data drop issue.

Popularity-Based Online Broker Selection Different from existing protocols, the online broker selection approach we proposed allows users to choose brokers *online* from his/her social connections (i.e., from the ones who are willing to contribute), while not requiring direct contacts with other users in the physical world. Users advertise their predicted popularity in the online community, and a publisher can choose the ones with highest popularity among them. Online broker selection also decreases the time cost on task allocation: the selected nodes can be allocated the dissemination task with no delay if they are online, while off-line brokers can be informed and obtain the allocated task once they are within an environment with Internet connection (hot spots, wired network, etc.)

Online Task Notification and Termination With the integration of online and off-line communities, the publisher can be notified about the matched node through online channels. The brokers can also be informed to terminate the dissemination task through online communication.

7.3.3.2   The HSN Infrastructure

We illustrate the infrastructure of HSN in Fig. 7.9. There are two kinds of components: *online components (blue colored)* and *opportunistic components (green colored)*. Online components are used for the linkage between mobile nodes and centralized online community server; opportunistic components run on the mobile phone and for opportunistic communication with other mobile nodes in the proximity.

Online Components

Request publication component allows users to post their new requests (buying things, making friends) to the server. Upon receiving a request, the *broker selection* component will choose k brokers from the online community. The *task allocation* component allocates the request to the k selected brokers. The brokers will carry and opportunistically disseminate the request to the nodes they contact. The *task notification* component notifies the matched request to its publisher. The request is terminated by its publisher after a successful transaction.

Opportunistic Components

With the *request duplication* component, the publisher and selected brokers copy the handling requests to the mobile nodes within the range of short-range radio. The *request matching component* matches the requests a node carries with new-coming requests.

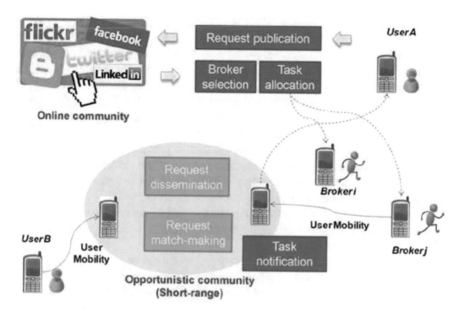

Fig. 7.9 The HSN infrastructure

7.3.4 Detailed Design of HSN

In this section, we present the main building blocks of HSN as well as the implementation of the HSN-based OBuy service.

7.3.4.1 Community Creation and Willingness-Based Broker Filtering

To address individual selfish based on social willingness, we have to extract the social relationship among users. There are many approaches that can be applied by HSN, such as extracting social connections from existing online social networks (e.g., Facebook) and predicting social ties among users to create new online communities. In this study, we focus on supporting user collaboration and information dissemination in urban environments (as demonstrated in the opportunistic trading use case), which raises a requirement to create city-dependent community. Note that we do not use friend lists that can be obtained from existing online communities, where "friends" may live far from each other (in different cities). Instead, we have proposed a contact-pattern-based community creation approach, which can derive social relations among users (as output) from historical user mobility traces (as input) to build an online community. The approach is described in detail below.

In [98], *contact time* (e.g., in the evening, at weekends) is used as an important predictor to identify interpersonal relation, where the ratio of encounters during non-working time is much higher for friends than non-friends. HSN has a broader definition about social relations, where we assume that people are willing to act as brokers for others with certain social ties, not merely friends. In other words, we do not distinct *family*, *friend*, or *colleague* relation, counting them whole as the *"familiar"* relation. We choose to characterize the familiar relation by means of two other important contact features: *contact duration* and *contact frequency*.

Contact duration. The duration of a contact is the total time that a tagged couple of mobile nodes are within reach of each other. Friends, family members, and colleagues usually have long contact durations. Compared to encounters that last for a certain time period, short encounters (e.g., two persons crossing each other in the street) are less important for the calculation of familiar relation. We introduce a threshold δ for contact duration, and only those encounter records that last longer than the threshold are used to estimate inter-user closeness.

Contact frequency. It is easy to understand that people who are familiar with each other usually have higher meeting frequencies. We use it as another predictor to identify the familiar relation.

The familiar relation is estimated leveraging historical encounter records. With the support of well-equipped mobile devices, long-term user meeting events can be tracked and recorded. However, the fact is that humans often alter their social behaviors and thus cause the change of social ties (e.g., starting a new job). Therefore, when calculating the familiar relation, we should give the historical encounter records a valid period VP. Encounter records expire *VP* are meaningless for the measurement of familiar relation. We use φ to represent the sum of valid encounters of two users (u_i and u_j) within the *VP* of encounter records, formulated as

$$\varphi(u_i, u_j) = N_{\text{VP}}(u_j), \tag{7.18}$$

where $N_{\text{VP}}(u_j)$ is the number of valid meeting times with u_j within VP period.

The familiar relation between u_i and u_j can be represented as

$$\text{BeFamiliar}(u_i, u_j) = \phi(u_i, u_j) > f, \tag{7.19}$$

where f is the contact frequency parameter.

With this, when choosing brokers for a request based on social willingness, we can filter the ones who do not have the familiar relation with the publisher to enhance reliable information dissemination.

7.3.4.2   Popularity-Based Online Broker Selection

We employ user popularity to measure a user's capability of acting as a broker. Each user's device continuously logs the devices it encounters and such encounter records are used to predict how many users this user is likely to meet in a forthcoming period ΔT.

We define this predicted value as $Pop_{\Delta T}(u_i)$. For example, $Pop_{7d}(u_i) = 25$ means that user u_i is expected to meet 25 different users in the next 7 days. We call $Pop_{7d}(u_i)$ as *weekly popularity* of u_i (simplified as $WP(u_i)$), which measures the number of users that u_i expects to meet in the next week. Similar to the familiar relation, the weekly popularity is also measured based on historical encounter records. We calculate the average value of u_i's historical weekly popularity values as $WP(u_i)$. All users advertise their predicted popularity in the online community, and a publisher can choose the ones with highest popularity among them. To advance information dissemination, people often need to choose multiple brokers rather than a single one [85]. The selected brokers can be informed and obtain the allocated task from the online community once they are within an environment with Internet connection.

7.3.4.3 Request Notification and Termination

Besides broker selection, there are two other problems to be addressed by HSN. They are how to notify the request publisher of the matched request and how to terminate brokers' work when the task is completed.

Matched result notification. If the matched node is found by the request publisher, the publisher can be directly informed. If it is found by a broker, the publisher should be notified by the broker. Two possible ways are available in HSN: (1) the broker can inform the information to the publisher through the online community (once he/she has Internet connection) and (2) an alternative notification method is asking the broker to communicate with the publisher by phone call or SMS once a matched node is found (more efficient than the prior method).

Task termination. The publisher resets the request state to "*completed*" if his/her request is matched. The updated information can be obtained by brokers when their mobile phone is online.

7.3.4.4 Use Case Implementation

Having described the main building blocks of HSN, in this subsection we present how the opportunistic trading service OBuy is implemented based on it.

There are two types of requests: sell request and buy request. To reduce network cost on data flooding, in OBuy, only sell requests will be disseminated, and the buyer (while not the seller) is notified when his/her request is matched. The buyer can select brokers and replicate his/her request to them. To lower traffic and communication cost, the brokers only carry and match the assigned requests but do not forward them to others. Based on the aforementioned broker selection and task management strategies, here we give the whole opportunistic trading algorithm. In the current stage, the algorithm is based on pre-collected human mobility traces.

The major components of HSN infrastructure are used here, including *request publication, broker selection, task allocation, request dissemination and match-*

making, and *task notification*. An example to showcase OBuy is illustrated in Fig. 7.10. *S1* publishes a sell request of "*Piano*" at time *T0*. *B1* publishes a buy request of "*Piano*" at *T1*. Two of *B1*'s social ties (*Br-m* and *Br-n*) are selected as brokers, and the tasks are allocated to them via online community at time *T2* and *T3*. When *S1* moves, people within the *n*-hop of him form an opportunistic community. Requests are shared and matched within the opportunistic community. At time *T5*, *S1* and *Br-n* meet in a coffee shop, and the buyer/seller requests are matched. *B1* is then informed of the matched result, through either online communication or phone calls. The task is terminated after the transaction is completed.

7.3.5   Performance Evaluation

To evaluate the performance of HSN, we designed a set of experiments based on the MIT Reality Mining dataset. For more information about the experimental setup, evaluation metrics, and baseline methods, please refer to the article [99]. The first experiment was designed to measure how broker popularity affects the performance of HSN. The popularity-based method (popularity in short) and the benchmark method random are used in the evaluation. The experiment process is described as follows:

(a) *Input specification: giving one buyer and a TTL, and generating a random seller.*
(b) *Repeat popularity and random 50 times, and compute the average success rate.*

Two buyers (69 and 79) with different number of "familiar" social ties are employed in the experiment. To investigate whether the functionality is affected by the user-tolerable matching latency, three *TTLs* (2, 7, and 14) are used, corresponding to *emergent*, *weekly*, and *long-lived trading need*. Results indicated that popularity-based method performs better than the random method, independent of the popularity of the node, the number of brokers k, and *TTL*. In addition, as we imagined, the success rate grows with the increasing of the number of brokers (i.e., k). However, the performance grows slowly after k is increased to a certain value. Moreover, with the growing of *TTL*, the success rate also increases. For *popularity*, it increases from 60 to 85% when *TTL* is extended from 2 to 7 ($k = 15$). This implies that emergent needs have low success rate, because short dissemination period decreases the opportunity to cover more nodes. By contrast, long-lived needs can be better served.

Second, to validate the effect of social selfishness on information dissemination, we have introduced the data drop rate in the traditional *off-line* method, which indicates the rate of task nonacceptance in terms of brokers' willingness. To make it simple, only one broker is used, who has to find one matched node (from three candidates in the community). The results are shown in Fig. 7.11. It can be seen that as the drop rate increases, the success rate (100 experiment times) also decreases.

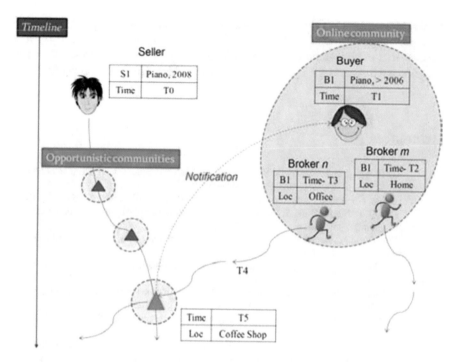

Fig. 7.10 Opportunistic trading: an example

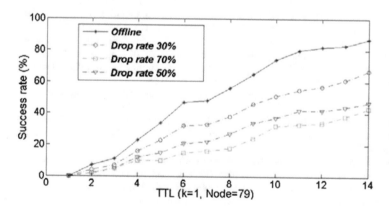

Fig. 7.11 The effect of social selfishness

This proved that data drop rate has a negative effect on opportunistic information dissemination, and thus our social willingness-based method is promising.

The last experiment was designed to validate the usefulness of hybrid social networking. We find that great performance improvement (in terms of success rate) is obtained when using HSN. For example, the improvement is about 15% for node

69 and almost 40% for node 79. On the other hand, the matching latency is largely reduced, decreasing about 60% for node 69 and more than 70% for node 79. It is because that the integration of an online community shortens the broker selection process, and increases the opportunity to select brokers with high popularity.

References

1. M. E. J. Newman and M. Girvan, Finding and evaluating community structure in networks, Physical Review E, 69, 26113–26127, 2004.
2. S. Fortunato, Community detection in graphs, Physics Reports, 486, 3–5, pp. 75–174, 2010.
3. Y.-Y. Ahn, J. P. Bagrow, and S. Lehmann, Link communities reveal multiscale complexity in networks, Nature, vol. 466, no. 7307, pp. 761–764, 2010.
4. J. D. Cruz, C. Bothorel, and F. Poulet, Entropy based community detection in augmented social networks. In CASoN. IEEE, 2011, pp. 163–168.
5. X. Wang, L. Tang, H. Gao, and H. Liu, Discovering overlapping groups in social media, in Proc. of ICDM'10, 2010, pp. 569–578.
6. I. S. Dhillon, Co-clustering documents and words using bipartite spectral graph partitioning, in Proc. of KDD'01. New York, NY, USA: ACM, 2001, pp. 269–274.
7. M.E.J. Newman, Modularity and community structure in networks, PNAS, vol. 103, pp. 8577–8582, 2006.
8. S. Scellato, C. Mascolo, M. Musolesi, and V. Latora, Distance matters: geo-social metrics for online social networks, in Proc. of WOSN'10. Berkeley, CA, USA: USENIX Association, 2010, pp. 8–8.
9. S. Scellato, A. Noulas, R. Lambiotte, and C. Mascolo, Socio-spatial properties of online location-based social networks. in Proc. of ICWSM' 11. The AAAI Press, 2011.
10. A. Noulas, S. Scellato, C. Mascolo, and M. Pontil, An empirical study of geographic user activity patterns in foursquare. in Proc. of ICWSM'11. The AAAI Press, 2011, pp. 570–573.
11. W. Chen, H. Yin, W. Wang, L. Zhao, W. Hua, and X. Zhou, Exploiting spatio-temporal user behaviors for user linkage, in Proceedings of CIKM'17, 2017, pp. 517–526.
12. Z. Cheng, J. Caverlee, K. Lee, and D. Z. Sui, Exploring millions of footprints in location sharing services. in ICWSM. The AAAI Press, 2011, pp. 81–88.
13. J. He, X. Li, and L. Liao, Category-aware next point-of-interest recommendation via listwise Bayesian personalized ranking, in IJCAI'17, 2017. pp. 1837–1843.
14. M.A. Vasconcelos, S. Ricci, J. Almeida, F. Benevenuto, and V. Almeida, Tips, dones and to dos: uncovering user profiles in foursquare, in Proc. of WSDM'12. New York, NY, USA: ACM, 2012, pp. 653–662.
15. N. Li and G. Chen, Analysis of a location-based social network, in Proc. of CSE'09. Washington, DC, USA: IEEE Computer Society, 2009, pp. 263–270.
16. A. Noulas, S. Scellato, C. Mascolo, and M. Pontil, Exploiting semantic annotations for clustering geographic areas and users in location-based social networks, in Proc. of ICWSM'11. The AAAI Press, 2011, pp. 32–35.
17. A. Clauset, M.E.J. Newman, and C. Moore, Finding community structure in very large networks, Physical Review E, 70, 66111–66116, 2004.
18. K. Wakita and T. Tsurumi, Finding community structure in mega-scale social networks, in Proc. of WWW'07. New York, NY, USA: ACM, 2007, pp. 1275–1276.
19. S. Cavallari, V. W. Zheng, H. Cai, K. C. C. Chang, and E. Cambria, Learning community embedding with community detection and node embedding on graphs, in Proceedings of CIKM'17, 2017, pp. 377–386.

20. V.D. Blondel, J.L. Guillaume, R. Lambiotte, and E. Lefebvre, Fast unfolding of communities in large networks, Journal of Statistical Mechanics: Theory and Experiment, 2008, 10, P10008, 2008.
21. G. Palla, I. Derenyi, I. Farkas, and T. Vicsek, Uncovering the overlapping community structure of complex networks in nature and society, Nature, vol. 435, pp. 814–818, 2005.
22. L. Tang and H. Liu, Community detection and mining in social media, Synthesis Lectures on Data Mining and Knowledge Discovery, vol. 2, pp. 1–137, 2010.
23. K. Steinhaeuser and N. V. Chawla, Community detection in a large real-world social network, in Social Computing, Behavioral Modeling, and Prediction, H. Liu, J. J. Salerno, and M. J. Young, Eds. Springer,New York, 2008, pp. 168–175.
24. M. Hosseini-Pozveh, K. Zamanifar, and A. R. Naghsh-Nilchi, A community-based approach to identify the most influential nodes in social networks, Journal of Information Science, 43, 2, 204-220, 2017.
25. Y. Zhou, H. Cheng, and J. X. Yu, Graph clustering based on structural/attribute similarities, Proc. VLDB Endow., vol. 2, no. 1, pp. 718–729, 2009.
26. L. Duan, W. N. Street, Y. Liu, and H. Lu, Community detection in graphs through correlation, in KDD'14, 2014, pp. 1376–1385.
27. M. McPherson, L. Smith-Lovin, and J. M. Cook, Birds of a feather: Homophily in social networks, Annual Review of Sociology, vol. 27, no. 1, pp. 415–444, 2001.
28. J. Cranshaw, E. Toch, J. Hong, A. Kittur, and N. Sadeh, Bridging the gap between physical location and online social networks, in Proc. of Ubicomp'10. New York, NY, USA: ACM, 2010, pp. 119–128.
29. L. Inc. Foursquare, About foursquare, April 2012. [Online]. Available: https://foursquare.com/about/
30. [Online]. Available: https://developer.foursquare.com/docs.
31. [Online]. Available: https://dev.twitter.com/docs.
32. M. Ye, K. Janowicz, C. Mülligann, and W.-C. Lee, What you are is when you are: the temporal dimension of feature types in location-based social networks, in Proc. of GIS'11. New York, NY, USA: ACM, 2011, pp. 102–111.
33. Z. Wang, D. Zhang, X. Zhou, D. Yang, Z. Yu and Z. Yu, Discovering and profiling overlapping communities in location-based social networks, IEEE Transactions on Systems, Man, and Cybernetics: Systems, vol. 44, no. 4, pp. 499-509, 2014. doi: https://doi.org/10.1109/TSMC.2013.2256890
34. M. Girvan and M. E. J. Newman, Community structure in social and biological networks, PNAS, 99, 12, 7821–7826, 2002.
35. Huynh, T., Fritz, M., Schiele, B. Discovery of activity patterns using topic models. In Ubicomp 10–19 (2008)
36. Liu, Y., Chen, L., Pei, J., Chen, Q., Zhao, Y.: Mining frequent trajectory patterns for activity monitoring using radio frequency tag arrays. In PerCom, pp. 37–46 (2007)
37. Zheng, Y., Chen, Y., Li, Q., Xie, X., Ma, W.Y.: Understanding transportation modes based on GPS data for web applications. ACM Trans. Web 4(1), 1–36 (2010)
38. Bonneau, J., Anderson, J., Anderson, R., Stajano, F.: Eight friends are enough: social graph approximation via public listings, In Proceedings of the Second ACM EuroSys Workshop on Social Network Systems, March 2009, 13–18 (2009)
39. Carley, K.M., Krackhardt, D.: Cognitive inconsistencies and non-symmetric friendship. Soc. Netw. 18(1), 1–27 (1996)
40. Vaquera, E., Kao, G.: Do you like me as much as I like you? Friendship reciprocity and its effects on school outcomes among adolescents.Soc. Sci. Res. 37(1), 55–72 (2008)
41. Yu, Z., Zhou, X., Becker, C., Nakamura, Y.: Tree-based mining for discovering patterns of human interaction in meetings. IEEE Trans. Knowl. Data Eng. 24(4), 759–768 (2012)
42. Eagle, N., Pentland, A., Lazer, D.: Inferring social network structure using mobile phone data. PNAS 106(36), 15,274–15,278 (2009)

43. Palla, G., Barabasi, A.-L., Vicsek, T.: Quantifying social group evolution. Nature 446(7136), 664–667 (2007)
44. Kumar, R., Novak, J., Tomkins, A.: Structure and evolution of online social networks. In Proceedings of 12th International Conference on Knowledge Discovery in Data Mining (KDD 2006), 611–617.
45. Musiał, K., Kazienko, P.: Social networks on the Internet. World Wide Web, online, doi: https://doi.org/10.1007/s11280-011-0155-z (2012)
46. Tang, L., Wang, X., Liu, H. Scalable learning of collective behavior. IEEE Trans. Knowl. Data Eng. online. https://doi.org/10.1109/TKDE.2011.38 (2011)
47. Onnela, J.-P., et al.: Structure and tie strengths in mobile communication networks. PNAS 104 (18), 7332–7336 (2007)
48. Chen, J., Saad, Y.: Dense subgraph extraction with application to community detection. IEEE Trans. Knowl Data Eng, online. https://doi.org/10.1109/TKDE.2010.271 (2011)
49. Lin, Y.-R., Chi, Y., Zhu, S., Sundaram, H., Tseng, B.L.: Analyzing communities and their evolutions in dynamic social networks. ACM Trans Knowl Discov Data 3(2), 8, 2009
50. Kossinets, G., Watts, D.J.: Empirical analysis of an evolving social network. Science 311 (5757), 88–90 (2006)
51. Malmgren, R.D., Hofman, J.M., Amaral, L.A.N., Watts, D.J.: Characterizing individual communication patterns. In KDD, pp. 607–616 (2009)
52. Leskovec, J., Horvitz, E.: Planetary-scale views on a large instant-messaging network. In WWW, pp. 915–924 (2008)
53. Hristova, D., Musolesi, M., & Mascolo, C.: Keep Your Friends Close and Your Facebook Friends Closer: A Multiplex Network Approach to the Analysis of Offline and Online Social Ties. In ICWSM. (2014)
54. Gonzalez, M.C., Hidalgo, C.A., Barabasi, A.-L.: Understanding individual human mobility patterns. Nature 453(5), 779–782 (2008)
55. Song, C., Qu, Z., Blumm, N., Barabasi, A.-L. Limits of predictability in human mobility. Science 327 (5968), 1018–1021 (2010)
56. Eagle, N.: Behavioral inference across cultures: using telephones as a cultural lens. IEEE Intell. Syst. 23 (4), 62–64 (2008)
57. Wesolowski, A., Eagle, N.: Parameterizing the Dynamics of Slums. In Proceedings of the AAAI Symposium on Artificial Intelligence and Development, pp. 103–108 (2010).
58. Dong, Z., Song, G., Xie, K, Sun, Y., Wang, J.: Adequacy of data for mining individual friendship pattern from cellular phone call logs. The 2009 Sixth International Conference on Fuzzy Systems and Knowledge Discovery, pp. 573–577.
59. Wang, H., & Li, Z.: Region representation learning via mobility flow. In Proceedings of the 2017 ACM on Conference on Information and Knowledge Management. pp. 237–246 (2017).
60. Zhang, J., Zheng, Y., & Qi, D.: Deep Spatio-Temporal Residual Networks for Citywide Crowd Flows Prediction. In AAAI, pp. 1655–1661 (2017).
61. Fan, Z., Song, X., Shibasaki, R., & Adachi, R.: City Momentum: an online approach for crowd behavior prediction at a citywide level. In Proceedings of the 2015 ACM International Joint Conference on Pervasive and Ubiquitous Computing. pp. 559–569 (2015).
62. Leenders, R.T.A.J.: Evolution of friendship and best friendship choices. J. Math. Sociol. 21 (1–2), 133–148 (1997)
63. Han, Y., & Tang, J. : Who to invite next? Predicting invitees of social groups. In Proceedings of the 26th International Joint Conference on Artificial Intelligence. pp. 3714–3720 (2017).
64. Khanafiah, D., Situngkir, H.: Social balance theory: revisiting Heider's balance theory for many agents. Technical Report, Bandung Fe Institute (2004)
65. Barabasi, A.-L., Jeong, H., Neda, Z., Ravasz, E., Schubert, A., Vicsek, T.: Evolution of the social network of scientific collaborations. Phys. A 311(3–4), 590–614 (2002)
66. Leskovec, J., Backstrom, L., Kumar, R., Tomkins, A.: Microscopic evolution of social networks. In KDD, pp. 462–470 (2008)

67. Liu, Y., Goncalves, J., Ferreira, D., Hosio, S., & Kostakos, V. : Identity crisis of Ubicomp?: Mapping 15 years of the field's development and paradigm change. In Proceedings of the 2014 ACM International Joint Conference on Pervasive and Ubiquitous Computing. pp. 75–86 (2014)

68. Cui, Y., Pei, J., Tang, G., Luk, W.-S., Jiang, D., Hua, M.: Finding email correspondents in online social networks. World Wide Web, online, doi: https://doi.org/10.1007/s11280-012-0168-2 (2012)

69. Hsu, C.W., Chang, C.C., Lin, C.J.: A practical guide to support vector classification. Technical Report, (2005)

70. Chang, C.C., Lin, C.J.: LIBSVM: a library for support vector machines. Software available at http://www.csie.ntu.edu.tw/~cjlin/libsvm (2001)

71. Heider, F.: The psychology of interpersonal relations. John Wiley and Sons, New York (1958)

72. Hallinan, M.T.: The process of friendship formation. Soc. Netw.1(2), 193–210 (1978)

73. Z. Yu, X. Zhou, D. Zhang, G. Schiele, C. Becker. Understanding social relationship evolution by using real-world sensing data. World Wide Web (2013) 16: 749. Doi: https://doi.org/10.1007/s11280-012-0189-x

74. B. Guo, Z. Yu, D. Zhang, X. Zhou, Opportunistic IoT: Exploring the Social Side of the Internet of Things, The 16th IEEE International Conference on Computer Supported Cooperative Work in Design (CSCWD'12), Wuhan, China, 2012.

75. M. Conti, M. Kumar, Opportunities in opportunistic computing, Computer, 43, 1, 2010, 42–50.

76. R. Grob, M. Kuhn, R. Wattenhofer, and M. Wirz, Cluestr: mobile social networking for enhanced group communication, Proc. of ACM GROUP, Sanibel Island, Florida, USA, 2009.

77. S.B. Mokhtar, et al., A Self-Organizing Directory and Matching Service for Opportunistic Social Networking, Proc. of the 3rd Workshop on Social Network Systems (SNS), Paris, France, 2010.

78. J. Kangasharju, J. Ott, O. Karkilahti, Floating Content: Information Availability in Urban Environments, Proc. of IEEE Percom'10, 2010.

79. N.D. Lane, et al., Exploiting Social Networks for Large-Scale Human Behavior Modeling, IEEE Pervasive Computing, 10, 4, 2011, 45-53.

80. A.T. Campbell, et al., The Rise of People-Centric Sensing, IEEE Internet Computing, 12, 4, 2008, 12-21.

81. Chiu, Chao-Min, et al. Understanding online community citizenship behaviors through social support and social identity. International Journal of Information Management 35(4) 2015: 504-519.

82. Kim, Jooho, and Makarand Hastak. Social network analysis: Characteristics of online social networks after a disaster. International Journal of Information Management 38(1) 2018: 86-96.

83. M. Motani, V. Srinivasan, P.S. Nuggehalli, PeopleNet: engineering a wireless virtual social network, Proc. of MobiCom'05, 2005.

84. D. Bottazzi et al., Context-Aware Middleware for Anytime, Anywhere Social Networks, IEEE Intelligent Systems, vol. 22, no. 5, 2007, pp. 23–32.

85. U. Lee, J.S. Park, E. Amir, M. Gerla, 'Fleanet: a virtual market place on vehicular networks, IEEE Transactions on Vehicular Technology, vol. 59, no. 1, 344-55, 2010.

86. W. Hsu, T. Spyropoulos, K. Psounis, A. Helmy, Modeling Time-Variant User Mobility in Wireless Mobile Networks, Proc. of InfoCom'07, 2007, pp. 758–766.

87. Alim, Md Abdul, et al. Structural vulnerability assessment of community-based routing in opportunistic networks. IEEE Transactions on Mobile Computing 15(12) 2016: 3156-3170.

88. Tao, Jun, et al. Contacts-aware opportunistic forwarding in mobile social networks: A community perspective. Wireless Communications and Networking Conference (WCNC), 2018 IEEE. IEEE, 2018.

89. Zhu, Konglin, et al. Data routing strategies in opportunistic mobile social networks: Taxonomy and open challenges. Computer Networks 93 (2015): 183-198.

90. J. Tang, T. Lou, J. Kleinberg, Inferring Social Ties across Heterogeneous Networks, Proc. of WSDM'12, 2012, pp. 743–752.

91. J. Cranshaw, et al., Bridging the gap between physical location and online social networks, Proc. of Ubicomp '10, Pittsburgh, PA, 2010.
92. D. Zhang, Z. Wang, B. Guo, V. Raychoudhury, X. Zhou, A Dynamic Community Creation Mechanism in Opportunistic Mobile Social Networks, Proc. of SocialCom 2011, MIT, USA, 2011.
93. T. Roughgarden, E. Tardos, How Bad is Selfish Routing? Journal of the ACM, 49, 2, 2002, 236–259.
94. Q. Li, S. Zhu and G. Cao, Routing in Selfish Delay Tolerant Networks, Proc. of InfoCom'10, 2010.
95. J.J. Jaramillo, R. Srikant, Darwin: Distributed and adaptive reputation mechanism for wireless ad-hoc networks, Proc. of MobiCom, 2007.
96. R. Ma, An incentive mechanism for P2P networks, Proc. of DCS, 2004, pp. 516–523.
97. M. Granovetter, The strength of weak ties, The American Journal of Sociology, vol. 78, no.6, 1973.
98. N. Eagle, et al., Inferring Social Network Structure using Mobile Phone Data, PNAS, vol. 106, no. 36, 2007, pp. 15274-15278.
99. B. Guo, Z. Yu, X. Zhou and D. Zhang, HybridSN: Interlinking Opportunistic and Online Communities to Augment Information Dissemination, 9th International Conference on Ubiquitous Intelligence and Computing and 9th International Conference on Autonomic and Trusted Computing, Fukuoka, 2012, pp. 188–195. doi: https://doi.org/10.1109/UIC-ATC.2012.29

Chapter 8
Open Issues and Emerging Trends

Abstract While there have been significant progress in the sensing and understanding of human behaviors, we still face a number of theoretical and technical challenges which need to be further explored. In this chapter, we discuss possible challenges and open issues from the aspects of human behavior itself, the data, as well as the models and evaluations, respectively. Afterwards, we present some emerging trends and directions, with hoped-for potential breakthroughs promising advanced human behavior sensing and understanding models and techniques.

8.1 Research Challenges

Human behavior sensing and understanding face many challenges, which can be summarized from the following three perspectives.

8.1.1 Challenges from Human Behavior Itself

Human behavior can be influenced by various factors, including culture diversity, different physical spaces, and multilevel social relations. Accordingly, there exist some characteristics of human behavior itself that are difficult to sense and analyze, e.g., capriciousness, evolution, and multiple granularity. Taking capriciousness as an example, one person's emotion can be easily affected by the behaviors of other people, while the influence factors are usually dynamic and uncertain. Therefore, capturing and quantifying these influence factors becomes a challenge. In addition, as human behavior ranges from fine-grained hand gestures to large-scale crowd movements, it could be another technical challenge to understand human behaviors in multiple granularity. For instance, understanding hand gestures requires

---

Part of this chapter is based on a previous work: Z. Yu, H. Du, F. Yi, Z. Wang, B. Guo. Ten scientific problems in human behavior understanding. CCF Transactions on Pervasive Computing and Interaction (2019) 1:3–9

© Springer Nature Singapore Pte Ltd. 2020
Z. Yu, Z. Wang, *Human Behavior Analysis: Sensing and Understanding*,
https://doi.org/10.1007/978-981-15-2109-6_8

techniques sensitive to capturing subtle changes, while for crowd behavior monitoring it is crucial to determine the global features from various individual behaviors.

8.1.2   Challenges from the Data

Not only because of the development of sensor-equipped mobile devices and online social network services, but also due to the dynamics of human behavior across different domains and spaces, there are unprecedented as well as diverse "Big Data" related to human behaviors. In specific, multiple interfaces (e.g., smartphones, mobile apps, and websites) are broadly used that can capture human behaviors with different formats (e.g., text, audio, and image) and various spatial-temporal information (e.g., different places and time slices). However, when researchers manage to conduct analysis on such plenty of datasets, they undoubtedly will face challenges during procedures such as data acquisition and data processing/analyzing. We identify the following key research challenges that may be involved, which contain data fragmentation, data heterogeneity, data representativeness, data sparsity, imbalanced data distribution, and spatial-temporal correlation. For example, due to the fact that population biases exist across different social media platforms [1], it becomes extremely difficult to gather human behavior data that could accurately and completely reflect/represent the real-world behavior, which hence causes the challenge of data representativeness. In addition, there also exists imbalanced data distribution problem since valuable data records may only have a small proportion in the whole dataset, which causes difficulty in modeling rare behavior patterns using such an insufficient dataset. In short, all the identified challenges are crucial to researchers, and we need to take proactive actions to deal with these challenges for better human behavior modeling and understanding.

8.1.3   Challenges from Modeling and Evaluation

There have been various models proposed for human behavior recognition and prediction, and these models usually regard the behavior computable as the default. However, human behavior is usually related to many dynamic and uncertain factors. Therefore, theoretical verification on computationality is essential yet difficult. From the aspect of evaluation, a valuable research should be comparable, whereas there is no standard evaluation metrics or systems as the context of human behavior varies. The determining of techniques applied, the baseline method, and the experimental environment are still to be further studied for building a standard of human behavior understanding. In addition, it is also crucial to determine what performance should be considered (e.g., accuracy, time efficiency, energy efficiency, robustness), which depends on the behavior characteristics. Taking human activity recognition (such as fall detection) as an example, firstly, recognizing the fall of older adults should be in time so that effective services can be provided. Second, the technique employed cannot be affected by the environments, which means that it should be robust. In brief, the evaluation for human behavior understanding makes a demand on the comprehensive consideration of behavior characteristics as well as the optimized model.

8.1.4   Ten Most Important Problems

Based on the challenges discussed above, we identify ten most important problems in human behavior understanding. The first three come from challenges related to human behavior itself, problems No. 4–8 are related to the data, and the last two are derived from the challenges of model and evaluation.

Behavior evolution: Human behavior can change over time; for example, a person preferring literatures could possibly turn to reading social science when he/she grows older. We call this characteristic behavior evolution, which requires the model constructed to adapt to such evolution as well as identify when the behavior changes. However, most existing models only output the results at present without any temporal/evolutionary details, which cannot support the detection of behavior evolution. To address this problem, there are several questions that should be considered. For instance, which types of human behavior can evolve? Can the applied model detect such behavior changes, including when and why they happen? Can the model adapt to the change and output time-variant results? Has the range of all possible results been known? Machine learning algorithms are widely used for human behavior understanding, especially pattern recognition, and existing studies usually utilize predefined features to identify one or multiple specific human behaviors. Apparently, such algorithms may not be suitable for the understanding of dynamic behaviors. A smart model is to be developed that can make adjustments adaptively, e.g., a multistage model [2]. More importantly, as we hope to reveal the law of human behavior, the descriptive information is expected to be obtained (e.g., why the pattern changes), which also helps decision-making.

Multi-aspect of human behavior: Multi-aspect refers that the appearance of a specific human behavior is not unique; that is, it shows difference in multiple aspects, such as human's personality. For example, a person can remain calm when communicating with colleagues, while he/she can be lively when facing the families. Moreover, there may also be some unknown appearances. As our understanding for a person is usually one sided, it is difficult to obtain an overall picture. As a result, many questions need to be addressed: For instance, what reflects the multi-aspect of human behavior? Can all aspects be understood? Which aspect can/cannot be sensed and analyzed? Is each aspect equal for characterizing human behavior? What are their weights? Facing these problems, we can first explore the method that can capture an overall portrait, and then make further analysis for each behavior aspect.

Capriciousness: Human behavior is capricious for the others and the environment. What factors can influence human behavior easily? Can these influences be quantified? As we know, a person's emotion could be affected by others, e.g., conformity psychology and emotional contagion. In most cases the influence is caused by subjective factors, the analysis of which needs to be combined with the knowledge in both psychology and sociology. Meanwhile, the influence is difficult to measure, and we are not sure whether all factors can be quantified. For the factors that are difficult to quantify, how can we estimate or model the influence? The solution may be studied by referring to similar models in other fields.

Data fragmentation: To capture a person's daily life activities, researchers have to apply various sensing devices that are distributed at separate places and time slots, which eventually leads to the fragmentation of the collected dataset. For example, the captured data can be distributed at different places (e.g., home, office, and gym) with different time logs (e.g., morning, noon, and night). Hence, knowing how to merge/aggregate the fragmented data together is necessary for understanding a comprehensive human behavior pattern, the exploration of which may need to consider the following questions. What are the relations among fragmented data? Which part of the fragmented data is important? According to different research problems, do we have to balance the weights across all the fragmented datasets? Or is there any criterion that could guide us for dealing with data fragmentation in a more effective way?

Data heterogeneity: Heterogeneity emphasizes the fact that the collected data are constituted by different formats, such as image, text, video, and audio. For example, a user's behavior can be tracked/recorded by various sensing devices (e.g., cameras, smartphones, and laptops) at different places and time slots, which results in a heterogeneous dataset. However, how we can collaboratively fuse all these different formats of data for human behavior understanding remains unsolved. For instance, when there is a parade happening in a city, it is usually expected that the event can be characterized by not only text-based news reports, but also pictures, videos, and user comments. However, it is a challenging issue to match all these heterogeneous data together for analysis: for example, how to discover the relationship between text-based contents and pixel-based images? How can we complement one type of data with other types of data? In what situations can the merging of heterogeneous data achieve better results than using singular data format? How can we balance the importance between various formats of data? Only if we could address these questions can our model be effective in human behavior modeling.

Spatial-temporal correlation: Location and time are two key features of human behaviors, and usually there are spatial and temporal correlations among the collected behavior digital footprints. For example, user's periodic activities could result in temporal correlated data records, and similar activities at similar places would produce spatial correlated data. Pan et al. [3] have proved the power of collaboratively exploring the spatial-temporal correlation between traffic information and online reviews. However, we still need to address a number issue to fully understand such spatial-temporal correlations: for example, in what ways we could match spatial and temporal data for problem-solving? Is the employed algorithm capable of taking data's spatial-temporal correlation into consideration during its learning process? If it has such a capability, then how can the spatial-temporal correlation constraint be formalized? Further, how can we evaluate the effectiveness of using spatial-temporal correlation features? Are there methods that can be used to adjust different weights across spatial and temporal factors? We also need to take care of fake correlations and over-usage of correlations. Some facts may seem to be correlated from the data, but indeed they are not. Moreover, correlations sometimes may be regarded as causality by mistake.

Data representativeness: The data collected may not be representative due to several reasons. (1) Population biases across different platforms: According to the investigation of Sprout Social,[1] we know that social media demographics vary a lot across different platforms, and such deviation may lead to misjudgments when applying knowledge from one platform to another. Therefore, we need methods and criterions to measure the similarity between population distributions of different platforms. Further, what sampling methods can be applied to collect a representative dataset? (2) Data protection from providers: It has been discussed in [2] that researchers are usually left in the dark about when and how social media providers change the sampling/filtering method of their data streams. Hence, the data available for researchers on the public platforms may be preprocessed by their providers due to privacy, safety, and other issues, which does not truly reflect users' real behavior patterns and might eventually affect the results of human behavior modeling. Therefore, we should focus on reasoning whether the obtained data is qualified for the problem or not, especially on modeling human behavior patterns.

Data sparsity: Although we are in the era of Big Data, we may still lack useful data for human behavior understanding. Meanwhile, researchers may be interested in user's abnormal/rare activities to fully characterize human beings. However, such data records usually only take a little proportion of the whole dataset, which causes difficulty in conducting this kind of research. Hence, how to deal with data sparsity becomes crucial. Although there are several off-the-shelf methods that can be applied to data sparsity problem, we still need to seek effective methods to handle imbalanced data distribution since those traditional methods are usually incapable of addressing such emerging challenges.

Computationality: Can all human behaviors be computed (including captured and modeled) by using the data and techniques available? How can the computationality be proved theoretically? In which conditions can they be computed? What limits the computationality? For example, Song et al. [4] explored the limits of predictability in human mobility by measuring the entropy of each individual's trajectory. Zhang et al. [5] proposed the Fresnel zone model to reveal the precision boundary of Wi-Fi-based activity recognition, which is different from traditional pattern-based behavior recognition approaches [6]. Inspired by these studies, we should first think about a fundamental question: To what extent is human behavior computable? Only based on the computationality can we improve the performance in human behavior

---

[1]https://sproutsocial.com/insights/new-social-media-demographics/

modeling, and also broaden the current field on human behavior understanding. This part still needs further exploration in theory.

Comparability: Is it possible to build a benchmark for human behavior understanding? Are there a set of pubic datasets for evaluation? Are there any techniques or algorithms that can be used as baselines? For specific behavior understanding, what are the basic evaluation criteria? And what is the numerical standard? For similar research problems (e.g., user positioning), researchers usually apply different sensing techniques, datasets, models, and algorithms. Moreover, the experimental environments are distinguishable. Thus, these studies cannot be compared completely because of these differences, and we cannot determine what result is acceptable, what result is perfect, and what result is unreasonable. It still lacks a complete evaluation system to highlight the experimental conditions and performance requirements.

8.2 Emerging Trends and Directions

8.2.1 Complex Behavior Recognition

Current work on activity recognition has mainly focused on simplified use scenarios involving single-user single-activity recognition. In real-world situations, human activities are often performed in complex manners. These include, for example, that a single actor performs interleaved and concurrent activities, multiple actors perform a cooperative activity, and/or a group of actors interact with each other to perform joint multiple activities. The approaches and algorithms described in previous sections cannot be applied directly to these application scenarios. As such, research focus on activity recognition has shifted towards this new dimension of investigation. We present some early work in this direction as follows.

In the modeling and recognition of complex activities of a single user, Wu et al. [7] proposed an algorithm using factorial CRF (FCRF) to recognize multiple concurrent activities. This model can handle concurrency but cannot model interleaving activities and cannot be easily scaled up. Hu and Yang [8] proposed a two-level probabilistic and goal-correlation framework that deals with both concurrent and interleaving goals from observed activity sequences. They exploited skip-chain CRFs (SCCRFs) at the lower level to estimate the probabilities of whether each goal is being pursued given a newly observed activity. At the upper level, they used a learnt graph model of goals to infer goals in a "collective classification" manner.

In the modeling and recognition of complex activities of group or multiple occupants, existing work has mainly focused on vision analysis techniques for activity recognition from video data. Various HMM models have been developed for modeling an individual person's behavior, interactions, and probabilistic data associations. These include the dynamically multilinked HMM model [9], the hierarchical HMM model [10], the coupled HMM [11], and the mixed-memory Markov model [12]. DBN models are also extensively used to model human interaction activities [13, 14], both using video cameras. Lian and Hsu [15] used FCRF to conduct inference and learning from patterns of multiple concurrent chatting activities based on audio streams. Work on using dense sensing for complex activity recognition is rare. Lin and Fu [16] proposed a layered model to learn multiple users' activity preferences based on sensor readings deployed in a home environment. Nevertheless, their focus is on learning of preference models of multiple users rather than on recognizing their activities. Wang et al. [17] used CHMMs to recognize multiuser activities from dense sensor readings in a smart home environment. They developed a multimodal sensing platform and presented a theoretical framework to recognize both single-user and multiuser activities.

While increasing attention has been drawn into this area, nevertheless, research in this niche field is still at its infancy. With the intensive interest in related areas, such as smart environments, pervasive computing, and novel applications, it is expected that research on activity recognition along this dimension will continue to receive attention and generate results in the next few years.

8.2.2   Multilevel Behavior Modeling for Scalability and Reusability

Current approaches and algorithms for activity sensing and understanding are often carefully handcrafted to well-defined specific scenarios, both activities and the environment. Existing implemented proof-of-concept systems are mainly accomplished by plumbing and hardwiring the fragmented, disjointed, and often ad hoc technologies. This makes these solutions subject to environment layout, sensor types and installation, and specific activities and users. The solutions, thus, suffer from a lack of interoperability and scalability. The latest experiments performed by Biswas et al. [18] indicated that it is difficult to replicate and duplicate a solution in different environments even for the same, simplest single-user single-activity application scenario. This highlights the challenge to generalize approaches and algorithms of activity recognition to real-world use cases.

While it is not realistic to predefine one-size-fits-all activity models due to the number of activities and the variation of the way activities are performed, we believe that multilevel activity modeling and corresponding inference mechanisms at different levels of details could address the problem of applicability and scalability. The basic idea is similar to the concepts of class and object in object-oriented programming (OOP) in that an activity is modeled at both coarse-grained and fine-grained levels. The coarse-grained level activity models are generic like a class in OOP that can be used by any one in many application scenarios. The fine-grained activity models are user specific like an instance in OOP that can accommodate the preference and behavior habits of a particular user. As such, the models and associated recognition methods can be applied to a wide range of scenarios. Chen and Nugent [19] developed activity ontologies where concepts represent the course-grained activity models, while instances represent user activity profiles. Okeyo et al. [20] extended this idea by developing learning algorithms to automatically create fine-grained individual-specific activity models, as well as learn new activity models to evolve ontologies towards model completion. Initial results are promising and further work is needed along this line.

8.2.3 Abnormal Behavior Recognition

Existing research on activity recognition focuses mainly on normal activities that may account for the majority of collected data and processing computation. Nevertheless, the results may contribute significantly less towards the purposes of activity recognition as most applications involving activity recognition intend to detect abnormal activities. This is a particularly important task in security monitoring where suspicious activities need to be dealt with and healthcare applications where assistances need to be provided for incapable users. While this view may generate cost-effective results, solving the problem is challenging. First, the concept of an abnormal activity has not been well defined and is elaborated with a variety of interpretations available. For instance, everyone performs activity A; one person carries out activity B. There are different views on whether or not activity B is abnormal. Yin et al. [21] defined abnormal activities as events that occur rarely and have not been expected in advance. Second, there is an unbalanced data problem in abnormal activity detection. Much larger proportion of sensing data is about normal activity, while the data for abnormal ones are extremely scarce, which makes training the classification model quite difficult. Knowledge-driven approaches can certainly fit in. The problem is really about the completeness of a priori domain knowledge. For example, is it possible to predict the behavior of a terrorist in advance or based on previous experience? Clearly, a raft of research problems and issues are open for further investigation.

8.2.4 Intent or Goal Recognition

Current activity recognition and its application is roughly a bottom-up approach starting from the lowest sensor data, and then discovering the activity and purposes of the user through increasing higher level processing. An emerging trend is to adopt a top-down approach to activity monitoring and recognition, namely to (1) recognize or discover the intent or goal of a user, (2) identify the activity that can achieve the goal, (3) monitor the user's behavior including the performed actions, (4) decide whether or not the user is doing the right thing in terms of the activity model and the monitored behavior, and, finally, (5) provide personalized context-aware interactions or assistance whenever and wherever needed. Goals can be either explicitly manually specified, such as when a care provider defines goals for a patient to achieve during a day, or learnt based on domain context. Activities are predefined in some flexible way and linked to specific goals. As such, once a goal is specified or identified, applications can instruct/remind users to perform the corresponding activity.

While research on cognitive computation, goal modeling, and representation of motivations, goals, intention, belief, and emotion has been undertaken widely in AI communities, in particular, within intelligent agent research, the adoption of the knowledge and research results in pervasive computing and smart environments and their applications have so far received little attention [22]. Nevertheless, interest is growing and a recent SAGAware has been organized [23], aiming to facilitate knowledge transfer and synergy, bridge gaps between different research communities/groups, lay down foundation for common purposes, and help identify opportunities and challenges.

8.2.5 Sensor Data Reuse and Repurposing

Currently, sensor data generated from activity monitoring, in particular, in the situations of using multimodal sensors and different types of sensors, are primitive and heterogeneous in format and storage, and separated from each other in both structure and semantics. Such datasets are usually ad hoc, lack descriptions, and are thus difficult for exchange, sharing, and reuse. To address these problems, researchers have made use of a priori domain knowledge to develop high-level formal data models. Nugent and Finlay [24] proposed a standard extensible markup language schema HomeML for smart home data modeling and exchange; Chen and Nugent [25] proposed context ontologies to provide high-level descriptive sensor data models and related technologies for semantic sensor data management aiming to facilitate semantic data fusion, sharing, and intelligent processing. We believe that knowledge-rich data modeling and standardization supported by relevant communities is a promising direction towards a commonly accepted framework for sensor data modeling, sharing, and repurposing. This idea is also in line with the infrastructure-mediated monitoring, namely deploy once and reuse for all. For more details about the emergeing trends and directions of human behavior analysis, please refer to the article [26].

8.3 Conclusion

Human behavior sensing and understanding is of great importance for various areas, ranging from personalized services and community-based recommendations to city-scaled planning, and other specific fields such as disease treatment and anti-terrorism. Considering the characters of human behavior itself, the behavior-related data, as well as the modeling and evaluation of human behaviors, we mainly outlined the ten most important scientific problems, and quite a number of open issues of human behavior sensing and understanding are not included. For example, we leave out specific modeling problems such as building a high-dimensional model, optimization algorithms for model learning, and combining advanced technologies like deep learning into existing models, which all play significant roles in human behavior understanding. We also present some emerging trends and directions, aiming to inspire constantly updated unsolved problems as well as solutions for the field.

References

1. A. Mislove, S. Lehmann, Y.-Y. Ahn, J.-P. Onnela, and J. N. Rosenquist, "Understanding the demographics of twitter users," ICWSM, vol. 11, p. 5th, 2011.
2. D. Ruths and J. Pfeffer, "Social media for large studies of behavior," Science, vol. 346, no. 6213, pp. 1063–1064, 2014.
3. B. Pan, Y. Zheng, D. Wilkie, and C. Shahabi, "Crowd sensing of traffic anomalies based on human mobility and social media," in Proceedings of the 21st ACM SIGSPATIAL International Conference on Advances in Geographic Information Systems. ACM, 2013, pp. 344–353.
4. Song, C., Qu, Z., Blumm, N., Barabási, A.-L.: Limits of predictability in human mobility. Science, vol. 327, no. 5968, pp. 1018–1021 (2010).
5. D. Zhang, H. Wang, and D. Wu, "Toward centimeter-scale human activity sensing with Wi-Fi signals," IEEE Computer, vol. 50, no. 1, pp. 48–57, 2017.
6. Z. Wang, B. Guo, Z. Yu, and X. Zhou, "Wi-Fi CSI based Behavior Recognition: from Signals, Actions to Activities," IEEE Communications Magazine, 2017.
7. T.Y. Wu, C.C. Lian, and J.Y. Hsu, "Joint recognition of multiple concurrent activities using factorial conditional random fields," in Plan, Activity, and Intent Recognition. Menlo Park, CA: AAAI Press, 2007.
8. D.H. Hu and Q. Yang, "Cigar: Concurrent and interleaving goal and activity recognition," in Proc. 23rd Nat. Conf. Artif. Intell., 2008, pp. 1363–1368.
9. S. Gong and T. Xiang, "Recognition of group activities using dynamic probabilistic networks," in Proc. 9th IEEE Int. Conf. Comput. Vis., 2003, pp. 742–749.
10. N. Nguyen, H. Bui, and S. Venkatesh, "Recognising behaviour of multiple people with hierarchical probabilistic and statistical data association," in Proc. 17th Brit. Mach. Vis. Conf., 2006, pp. 1239–1248.
11. N. Oliver, B. Rosario, and A. Pentland, "A Bayesian computer vision system for modeling human interactions," IEEE Trans. Pattern Anal. Mach. Intell., vol. 22, no. 8, pp. 831–843, Aug. 2000.
12. T. Choudhury and S. Basu, "Modeling conversational dynamics as a mixed memory Markov process," in Advances in Neural Information Processing Systems. Cambridge, MA: MIT Press, 2005, pp. 281–288.
13. Y. Du, F. Chen, W. Xu, and Y. Li, "Recognizing interaction activities using dynamic Bayesian network," in Proc. Int. Conf. Pattern Recognit., 2006, pp. 618–621.
14. D. Wyatt, T. Choudhury, and H. Kautz, "A privacy sensitive approach to modeling multi-person conversations," in Proc. 20th Int. Joint Conf. Artif. Intell., 2007, pp. 1769–1775.
15. C.C. Lian and J.Y. Hsu, "Chatting activity recognition in social occasions using factorial conditional random fields with iterative classification," in Proc. 23rd AAAI Conf. Artif. Intell., 2008, vol. 3, pp. 1814–1815.
16. Z.H. Lin and L.C. Fu, "Multi-user preference model and service provision in a smart home environment," in Proc. IEEE Int. Conf. Autom. Sci. Eng., 2007, pp. 759–764.

17. L. Wang, T. Gu, X. Tao, and J. Lu, "Sensor-based human activity recognition in a multi-user scenario," European Conference on Ambient Intelligence, Springer-Verlag, LNCS vol. 5859, pp. 78–87, 2009.

18. J. Biswas, A. Tolstikov, C. Nugent, L. Chen, and M. Donnelly, "Requirements for the deployment of sensor based recognition systems for ambient assistive living," in Proc. Int. Conf. Smart Homes Health Telematics, 2010, pp. 218–221.

19. N. Yamada, K. Sakamoto, G. Kunito, Y. Isoda, K. Yamazaki, and S. Tanaka, "Applying ontology and probabilistic model to human activity recognition from surrounding things," IPSJ Digital Courier, vol. 3, pp. 506–517, 2007.

20. G. Okeyo, L. Chen, H. Wang, and R. Sterritt, "Ontology-based learning framework for activity assistance in an adaptive smart home," in Activity Recognition in Pervasive Intelligent Environments (Atlantis Ambient and Pervasive Intelligence), vol. 4. Amsterdam, The Netherlands: Atlantis Press, 2011, pp. 237–262.

21. J. Yin, Q. Yang, and J. Pan, "Sensor-based abnormal human-activity detection," IEEE Trans. Knowl. Data Eng., vol. 20, no. 8, pp. 1082–1090, Aug. 2008.

22. D.H. Hu, S.J. Pan, V.W. Zheng, N. Liu, and Q. Yang, "Real world activity recognition with multiple goals," in Proc. Tenth Int. Conf. Ubiquitous Comput. (Ubicomp 2008), 2008, pp. 30–40.

23. L. Chen, I. Khalil, Z. Yu, W. Cheung, and P. Rashidi. (2011). Situation, activity and goal awareness (SAGAware2011), in Proc. SAGAware Workshop.

24. C.D. Nugent and D.D. Finlay, "HomeML—an open standard for the exchange of data within smart environments," in Proc. 5th Int. Conf. Smart Homes Health Telematics, 2007, pp. 121–129.

25. L. Chen and C.D. Nugent, "Semantic data management for situation-aware assistance in ambient assisted living," in Proc. 11th Int. Conf. Inf. Integr. Web-Based Appl. Services, 2009, pp. 296–303.

26. Z. Yu, H. Du, F. Yi, Z. Wang, B. Guo. Ten scientific problems in human behavior understanding. CCF Transactions on Pervasive Computing and Interaction (2019) 1:3–9.

Printed in the United States
By Bookmasters